C000170873

MOON POWER

A
CONSCIOUS
GUIDE

MERILYN KESKÜLA-DRUMMOND

MOON POWER

Empowerment through
cyclical living

aster

Dedicated to Grandmother Moon

An Hachette UK Company
www.hachette.co.uk

First published in Great Britain in 2020 by Aster, an imprint of
Octopus Publishing Group Ltd
Carmelite House
50 Victoria Embankment
London EC4Y 0DZ
www.octopusbooks.co.uk

Layout and Design Copyright © Octopus Publishing Limited 2020
Text Copyright © Merilyn Kesküla-Drummond 2020

Distributed in the US by
Hachette Book Group
1290 Avenue of the Americas
4th and 5th Floors
New York, NY 10104

Distributed in Canada by
Canadian Manda Group
664 Annette St.
Toronto, Ontario, Canada M6S 2C8

All rights reserved. No part of this work may be reproduced or utilized in any form
or by any means, electronic or mechanical, including photocopying, recording or by
any information storage and retrieval system, without the prior written permission
of the publisher.

Merilyn Kesküla-Drummond has asserted her right under the Copyright, Designs
and Patents Act 1988 to be identified as the author of this work.

ISBN 978-1-78325-340-1

A CIP catalogue record for this book is available from the British Library.

Printed and bound in China

10 9 8 7 6 5 4 3

Consultant Publisher: Kate Adams
Senior Editor: Pollyanna Poulter
Art Director: Juliette Norsworthy
Production Manager: Lucy Carter
Copy Editor: Jane Birch
Proofreader: Alison Wormleighton
Indexer: MFE Editorial
Illustrator: Sosha Davis @soshacreates
Designer: Rosamund Saunders

The information given in this book is not intended to act as a substitute for
medical treatment.

Contents

PART 1
Introduction

Why the Moon? Why now?

I'm glad you're here. I'm so grateful for you. You, who know there's more to life than permanent and linear grind. Also I'm grateful for all the teachers – spirit or living – who have whispered the contents of this book into my ear, and all the people around me who have facilitated this dream coming true.

Maybe it's the deep healing of the feminine and masculine that's happening all over, or the growing desire to interweave ancient practices with modern hi-tech, or our need to reconnect with nature and the cosmic realms, or the focus on women's empowerment, or anything else currently floating in our collective consciousness, that has reignited our fascination with the Moon – or maybe because, while writing this book, we also hit the 50-year mark from the Moon landing in 1969.

The bottom line is that our world is interconnected. As we are waking up from the long slumber of industrialism and patriarchy, the serpent of our individual soul power is raising its head. There's a deepening respect towards women and the feminine creative power and more gratitude for Mother Earth, the soil and all beings around us.

We are called to align our breath with the Universal breath and our voices with the voices of the Earth and the stars.

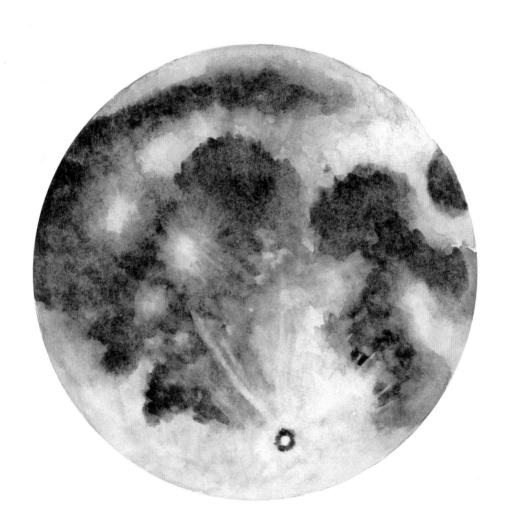

Where to start?

With all the healing modalities, therapies and retreats available, how do you know where to start your own journey?

I believe that each of us has a personal blueprint for reaching optimal performance. A map to lead us towards success, based on flow and sustainable growth. Each of us is part of the Universal rhythm, but also distinctly unique and individual. The map I want to give you for this journey is the one of *la Luna,* the beautiful waxing and waning ally of the Earth. With the Moon, you can learn to align and sustain your energy as you flow with the cosmic river of divine timing, reach your goals and identify the optimal times for expansion and retreat.

If you think about it, *everything* is cyclical: the way our bodies work – such as our respiratory or reproductive systems, or our heart beat – the movement of planets, the constant flow of day and night, and the seasons. All of these cycles affect our life in significant ways. We can either stubbornly march on the old path, or pull back and observe, understand the inner workings of these cycles, their rhythmical breath, trust their ways and align our lives with the pattern that nature gave us. And, as a consequence, develop more ease, creativity and flow.

This is the access-all-areas pass for your inner authority.

The bear never ponders, when it gets to late autumn: 'You know what, I'm not going to hibernate this winter. It's just not productive! I'd better keep going throughout winter to get there [where?] quicker.' Our human species seems to be the only one with a belief that we can go against nature and our physical and energetical flow, set at the time we were born. It's impossible to just inhale, we also need to empty our lungs, together with the carbon dioxide that is released by our bodies. It seems, though, that in modern times we have literally (and metaphorically) forgotten how to exhale. Do you hold your breath while you are composing an email?

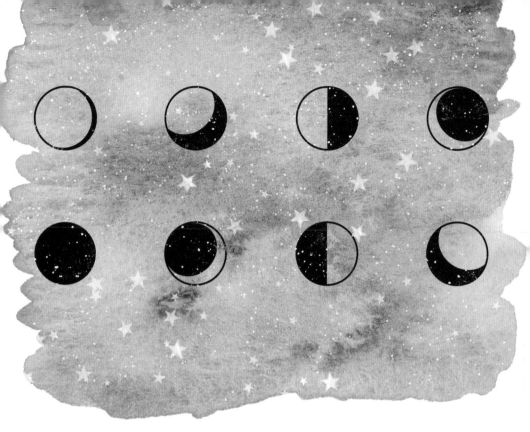

THE MOON HELPS YOU TO:

☽ Fuel your radiance by setting healthy boundaries.
☽ Rest and play more.
☽ Know when to schedule time for envisioning, planning or reflecting.
☽ Understand your needs and learn the tools to get them met, and so much more.

Natural cycles – like the menstrual cycle, Moon phases and seasons – are the ancient blueprints for your body and soul, providing the map for your divine timing and success.

If you're done struggling through yet another day filled with exhaustion, stress and anxiety while staring at your mile-long to-do list, then I invite you to embrace cyclical living. This is the ultimate time to transform the way you live your life, build your business or birth anything new into this world. It is an invitation to start observing and understanding the cyclical patterns within and around you. This is your map to inner strength and creative fertility, by honouring your inner authority, natural pace and divine timing.

MAPPING THE MOON, WHEREVER YOU ARE:

While this book has been written from a northern hemisphere perspective, the information given is perfectly relevant if you live in the southern hemisphere. The appearance of the Moon in the southern hemisphere is simply the mirror image of how the Moon appears in the northern hemisphere, and the phases are the same. So, in the northern hemisphere, the moon waxes from the right, while in the southern hemisphere it waxes from the left.

How I became a Moon advocate

Mylky Moon Lab...this was the name that came to me some years ago, completely out of the blue, like a download, a hint from the Universe, with no further explanation about what it should be or become. This happened after a decade of something I could describe as half-drowning in ice-cold water – when your body seems numb and half-frozen, disconnected, and you're constantly flapping around struggling to get your head to the surface for a breath of air.

As an event producer and creative marketing specialist for some of the well-known brands in fashion and tech, you wouldn't really think that, in essence, I was totally disempowered. But with constant highs and lows and an inability to set boundaries that would strengthen my own core being, I felt like the lines between me, other people, my work, expectations and the environments I was in were always blurred. I felt like a small boat being tossed around in a storm. I had a constant sense of unease, feeling unsafe and without an anchor to ground me.

I did not thrive in this environment, but my experiences did help me to see that our current working culture has also become disconnected from any sense of natural cycles, from feminine energies such as receptivity and, as a consequence, many of us are running on empty.

So, instead of keeping up with the pace I myself had created, and feeling like a failure if I ever got physically and mentally exhausted, I *paused* (which is quite a luxury these days) and began a deeper and committed journey of self-discovery. This translated into Mylky Moon Lab – an exploration of finding one's rhythm for empowered living and creating by honouring cyclical ebb and flow instead of pursuing a linear path.

> When we tune into the Universal patterns that surround us, while also acknowledging our own internal cycles, we create new paradigms for how we live and work.

We need to drop into a place of intuition, receptivity and body wisdom in order to start creating and living from a different, more aligned space.

Therefore, Mylky (= nourishment) Moon (= phases and patterns) Lab (= research, exploration, innovation) is not a ready-made product, but an ever-evolving being, an invitation to explore living in alignment with the natural cycles and patterns within and around us. It is also an invitation to keep creating our lives and businesses from an abundant and nourished foundation and not from a state of fatigued disempowerment.

I wasn't born in a hippie commune, wrapped in cosmic love, but had quite a disturbed childhood in Soviet Estonia. I wasn't introduced to meditation and yoga at an early age. I never travelled to India to follow a guru, I drank a fair share of alcohol, spent a night in a police cell after passing out on the street, hated sex, but had a lot of it, and spent half of my time on airplanes for the best part of a decade. I was lucky enough to listen when I needed to pause and to discover a deep knowing that, in order to build a sustainable society that gives back as much as it takes, we need to switch from the constant pursuit of growth to the cyclical model of expansion and contraction – similar to inhale and exhale. We need to respect the yin – being receptive and slow – as much as the yang – doing and action. And this journey starts from within.

MYLKY M^C ⟩N LAB

Why Moon power?

This book is to open the door for you. To help you explore 'the calendar of the Moon' and how to live and create aligned by its energetic shifts. This is the path of radical self-care.

Mainstream astrology, such as the horoscopes you might quickly read in a daily newspaper, is based on looking at Sun signs only. The Sun is traditionally associated with a masculine, external point of view – how we are going to shine and how we are seen by others.

The Moon, on the other hand, is shy and hidden compared with the beaming ball of fire that is the Sun. Uninterested in playing the lead role, she is more like a backstage worker in the theatre, pulling the strings, raising and lowering the curtain and making sure that the right actors are ready and on stage when they need to be. The Moon is the one who is the key for the story to emerge. If there were only the Sun – the lead actor – there would be no coherent and meaningful play, no unfolding story.

So, to step back to our own personal experience – what does this mean? It means that the Moon is a mirror for our soul path, the one that governs all our experiences – joyful, sad or scary – and all our emotions. The Moon is our daily life and the way we should live it moment by moment in order to get us

nearer to our purpose and manifest the gifts we are here to share. While the Sun is about light (and the 'end goal'), the Moon also deals with the shadow.

Even if you look at it from a purely astronomical standpoint, the Moon is the closest celestial body to us and the creator of the rhythmical ebb and flow, pull and push that affect the waters on the Earth. The Moon also lights up the night sky – a symbol of our soul, subconscious or dream state – and when it's fully illuminated (as during the Full Moon), many of us feel a strong soul activation and heightened emotions. Additionally, geological samples from the lunar surface show some similarities to those from the interior of Earth – so *la Luna* is mirroring our inner landscape in more ways than one.

The truth is we are cyclical beings, breathing together with the Universal rhythm, affected by the change of seasons, day and night and movement of planets in our cosmic galaxy.

By tuning into the inherent wisdom of our own bodies and natural patterns, through reconnection and alignment with the cyclical ebb and flow within and around us, we can rebuild our life and work on a sustainable foundation, filled with creativity, intuition and ease.

Maybe you have been wondering why you feel full of energy, social and confident one day and withdrawn, sensitive and emotional another. Why on some days everything seems to flow and other times you wake up in the morning 'not feeling yourself' and nothing seems to go as planned. Energetic, fatigued or creative – our physical energy and emotions are in constant change, in tune with the internal cycles and the cosmic patterns. Learning to work with these patterns can have profound effects on every aspect of your life: work, relationships and health. When you are able to observe and anticipate your physical and mental states, you can shape your schedule accordingly.

I wrote this book to invite you to tune into the Universal patterns that surround you, while also acknowledging your own internal cycles – this is how we'll create the new paradigm and reconnect to living and working in alignment with nature. Learn to follow your flow. Learn to honour nature, night, dreams, magic, creativity and subconscious, fertility and femininity as well as light, speed, action and the conscious mind.

The description and rituals that you'll find on the following pages are inspired by what I've learned about the Moon through exploring astrology, from books to courses, personal experience of observing the Moon's journey in relation to my daily life, and intuitively channelled information – from the Moon Letters I've been publishing bi-monthly for a few years now.

What is your Moon intention?

Why did you pick up this book?

Which part of your life are you most eager to change or heal?

What is your number one goal for the next year or month?

I invite you to formulate YOUR MOON INTENTION. One key aspect you'd like to work on - like a promise to yourself. Write it down in your notebook or Moon Diary (see page 72) or on a sticky note that you can put inside your bathroom cabinet. This will remind you WHY you are committing the time and energy to learn more about our lunar friend and the cosmic mysteries.

Here are some examples:

- ☽ My intention is to radically re-think and re-design all processes, to integrate cyclical perspective, and kick-start a powerful shift in the way I dream, create, develop and transform my life/career/business.
- ☽ I deepen my awareness of my physical and mental energy and start honouring my ever-changing nature.
- ☽ I honour the flow of creativity, efficiency and ease in my way of working – hustling, busyness and burnouts get kicked off the pedestal!
- ☽ I identify my needs and desires and establish ways to have them met.
- ☽ I learn the tools to navigate relationships better and understand that no one's perfect (myself included).
- ☽ I build a natural rhythm into my work and create a schedule that fits in with my life and my energy.
- ☽ I reconnect with a more ancient and natural way of living, which I already know in my bones. I will live in synchronicity with the natural world that is all around me and honour the divine timing.

You can also add a timeframe to give your intention more drive, for example:

In _____ months I will:

- ☽ Trust my own body and honour my inner authority over anything external.
- ☽ Have set my own pace and daily stress will be a thing of the past.
- ☽ Have radically slowed down – resting more, playing more – while actually becoming more creative, productive and successful.
- ☽ Have an intimate relationship with my unique flow and blueprint, unfolding my path to my destination (success!).
- ☽ Respect myself (my time and energy), which makes others respect me (and my work).
- ☽ Be able to harness the skills of my colleagues and team by understanding the energy available at any given time.

Write your Moon intention below. Most important though: with this kind of work, there are no rules, but YOUR rules. Everything here is meant as invitation and inspiration.

....

....

....

....

....

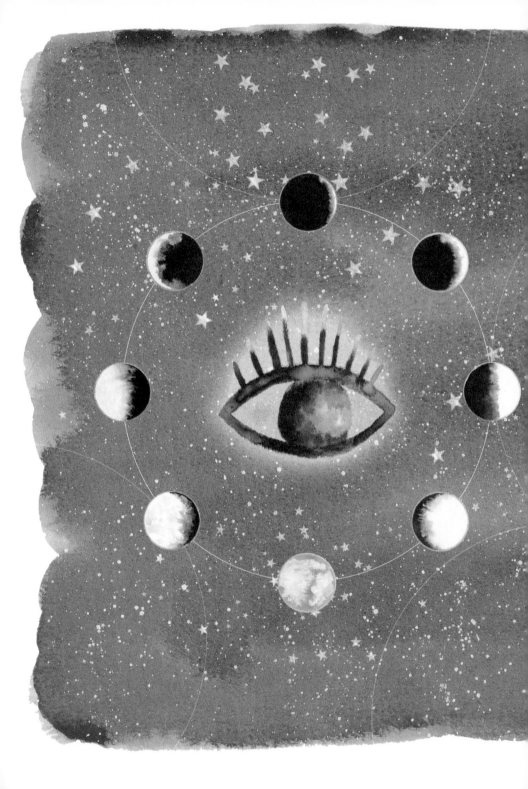

PART 2
Who is Moon?

Where did the Moon come from?

The Moon is never static and still, but a dynamic entity with many different facets – just like us. She's constantly in motion, circling around the Earth like a loyal companion, journeying through star constellations and forming aspects with other celestial bodies on the way. Without the Moon stabilizing the Earth's orbit, it is thought that our climate would be much more extreme. But how did the Moon come into existence?

There used to be a number of theories about how the Moon came to be. Until recently, the most widely accepted of these was the giant-impact theory, which says that the Moon was formed during a collision between the Earth and a smaller planet called Theia. During this huge crash in space, most of the Earth and Theia melted and formed as one, with the debris breaking off and creating the Moon.

This theory, however, has its challengers as most models suggest that more than 60 per cent of lunar material should be made up of the parts of Theia, when in fact the composition of the Earth and Moon is isotopically similar, so there is no trace of Theia. This has made the scientists question the Theia theory and brought about the multiple-impact hypothesis – which holds that the Earth collided with several bodies, rather than just one, sending clouds of rubble and vapour from Earth into orbit to form small moons that eventually all joined up – as a new way of explaining Moon's birth.

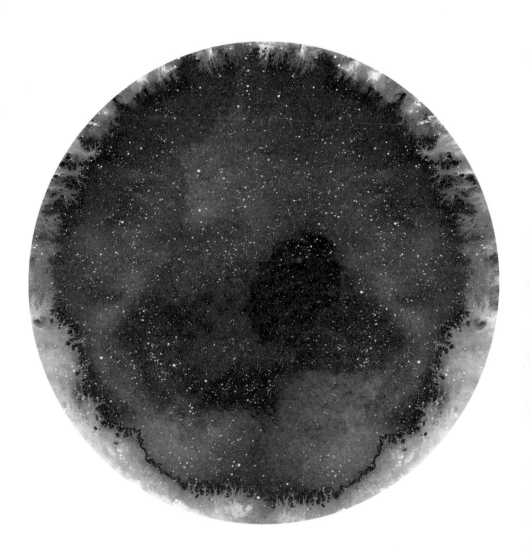

What is the Moon made of?

I always liked the idea that the Moon is made of cheese. Kind of cute, right? Actually, the Moon is much 'cooler' than that.

Like Earth, the Moon has a crust, a mantle and a core that is believed to be of solid iron, wrapped in an outer core of softer, liquid iron. The inside of *la Luna* is called the lithosphere, a layer about 1,000 kilometres (620 miles) thick, where magma formed, later creating the lava plains on the Moon's surface. There is no more volcanic activity on the Moon though, and, from a geological point of view, the Moon is inactive.

Scientists used to think that the dark patches on the Moon's surface were oceans, so they called them *maria* (singular, *mare*), Latin for 'seas'. In fact, these are 'seas' of hardened lava.

The Moon's crust is covered with regolith – a layer of loose debris covering solid rock, made of dust, soil or broken rock. This layer covers the entire Moon in uneven patches. The lighter-coloured areas are lunar anorthosites, igneous rock formations. The darker areas – the lunar 'seas' – are made of basalt (dark volcanic rock). When we look at the Moon, it resembles a face made out of its 'watery' features, eyes formed of Mare Imbrium (Sea of Rains) and Mare Tranquillitatis (Sea of Tranquillity) and the mouth of Mare Nubium (Sea of Clouds). There are many different myths about the shadows on the Moon and what they're thought to represent (see page 31).

All the Moon rocks brought back over the years have been dated by radiometric dating techniques and range in age from about 3.16 billion to about 4.44 billion years old. To compare, the oldest rocks from the Earth are between 3.8 and 4.28 billion years old. This would confirm the idea that the Moon and the Earth were formed at approximately the same time.

The Space Race

In the 1950s and 60s, the Moon became the desired destination, especially for Russia and the USA, getting the whole world involved in the rivalry called the Space Race.

While the USA was open and proud about its achievements, Russia was much more secretive and propaganda-based, shining light on their wins and keeping information about the challenges and failures of experiments to themselves. The competition between the major global players in the Space Race was about more than just space exploration and Moon landings, it was a symbol of the countries' strength, determination and perseverance – almost a measure of their power.

Soviet Space dogs became a major national obsession. They were mostly female stray dogs, picked up from the streets of Moscow and sent to space in specially customized modules to test the feasibility of space travel. The dogs were highly trained in wearing space suits, standing still for long periods of time and riding in centrifuges that simulated rocket launch. Obviously it raises many questions about animal rights, but in the 1950s/60s Soviet Union, these dogs – space explorers – were heralded as proletarian heroes. Pictures of the famous dogs – such as Laika, the 'barker', who died far away from

Earth, from panic and heat exhaustion, but was said to be 'alive and well' post-mission by the Soviet media – decorated a range of everyday objects, from cigarette cases to plates, clocks and postcards, and the animals became cult figures.

The success of America's Apollo 11 Moon landing in 1969 was largely thanks to some rather bright female engineers. Margaret Hamilton, who is now in her 80s, led NASA's team of software engineers and helped to develop the onboard flight software for the Moon missions. Thanks to her approach, there was not a single software bug on the Apollo flights. Katherine Johnson gained the reputation of NASA's 'human computer' and calculated the trajectories, launch windows and emergency back-up return paths for the space flights, including that of Apollo 11.

'We came in peace for all mankind'

– plaque left on the Moon by Apollo astronauts, July 1969

The dance between the Sun, Moon and Earth

The Moon reflects about 12 per cent of the sunlight that hits its surface. Throughout the book, when we talk about the Moon phases and movement through star constellations, we take the point of view of the Earth inhabitant and analyse the continuous dance between three bodies – the Sun, Moon and Earth.

The Moon is about 384,400 kilometres (238,855 miles) away from the Earth, locked into its elliptical orbit by gravity. Earth orbits the Sun at an average distance of 149,597,870 kilometres (about 93 million miles) – a unit of length called an astronomical unit, or AU.

Although the Sun is almost 400 times bigger than the Moon, it is also almost 400 times farther away – this is what makes the solar eclipse possible.

The gravitational pull by the Moon and the Sun creates two low and two high tides daily, occurring in specific places on Earth, depending on the rotation of our planet. As the Moon is so much closer than the Sun, it's the main culprit for pushing and pulling our waters. The tides actually have the power to change the speed of both the Moon and the Earth – the Moon is slightly pulled ahead by the rising water while the pace of the rotating Earth is slowed down. This means that our days are becoming longer, although just by a tiny fraction of a second per century.

Adult humans have around 70 per cent of fluid in their bodies, most of it in the fluid within cells. So does the Moon's gravity also 'dance' with our physical body?

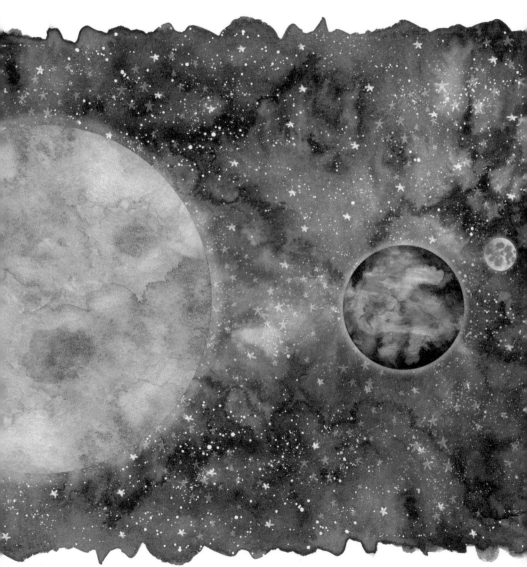

Space Age fashion

In fashion, the Space Age of the 1960s marked a pivotal change in the aesthetic and use of materials. The 'new look' featured monochrome palettes, sharp lines and prints of space objects or textile patterns resembling the surface of the Moon.

The godfathers of this new cosmic style were the French designers André Courrèges and Pierre Cardin, and Spanish-born designer Paco Rabanne. Courrèges' Space Age look included goggles and helmets, opaque sunglasses and silver trousers. In addition to his love for geometric shapes, he was also the first to incorporate plastics and PVC into clothing. His so-called Moon Girl look featured clean silhouettes and mostly clinical white outfits, with some vibrant oranges and yellows.

Rabanne created his own 'Space Age armour' of metal chain-mail minidresses, which were seen as 'uniforms for female empowerment'. He was consequently asked to create the look for Barbarella, the eponymous role played by Jane Fonda, in Roger Vadim's 1968 cult film.

Pierre Cardin was actually the only civilian to try on a NASA space suit. In 1968 he designed his renowned Cosmocorps collection and invented Cardine – an early tech material that was heat-treated to hold embossed designs. Cardin also designed *robes électroniques* – clothes decorated by light-up LED embroideries.

Lunar inspiration seeped everywhere and seemed to touch everything. In London, fashion designer Mary Quant used plastics in her Op Art-inspired, white-and-silver rainwear collection 'Wet'. 'Moon Boot', a snow boot designed by Giancarlo Zanatta, copied the shape and technology of astronauts' boots and, in music, David Bowie created his alter ego, astronaut Major Tom, for 'Space Oddity'. His first single to chart in the UK, it reached the top five on its initial release.

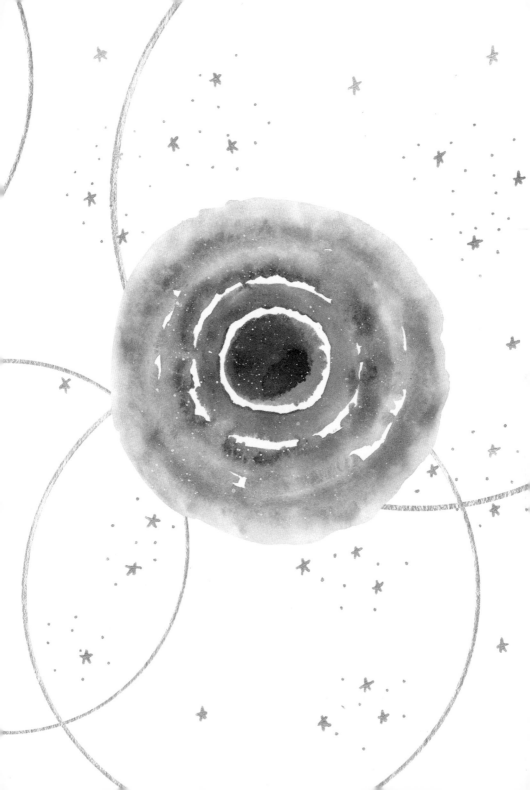

The Moon in different cultures

The Moon holds a myriad of names and is represented in countless stories since ancient times.

Moon always had her own deities – mostly goddesses who represented different phases of the Moon with female archetypes. New Moon, the time of youthfulness, hope and vitality, was symbolized by the Roman Moon goddess, Diana – a maiden huntress and the guardian of new life, a free spirit who roamed the woodlands and protected young girls and animals. In Greek mythology, Diana's equal is the hunting goddess Artemis who represents the wild and youthful aspect of the Moon.

Full Moon deities are the symbols of fertility, abundance and motherhood. Greek goddess Selene, sister to Sun god Helios, bore 50 daughters and her name is the root of the word 'selenology' – the scientific study of the Moon. Isis

is one of the most important Egyptian goddesses, the Giver of Life, governing marriage, fertility and childbirth. And Rhiannon, the Celtic Moon goddess, symbolizes rebirth, wisdom, transformation and artistic inspiration.

The waning and dark Moon was depicted by old, wise – often wild and destructive – female goddesses. These include the much feared dark Lilith, the Hindu goddess Kali and the Ancient Greek Hecate. They bring rebirth through healing and the breaking down of old structures.

Although female lunar personification is more well known, there are also many male gods associated with lunar powers, especially in cultures outside of the Western world – Sin in Mesopotamia, Thoth, Konsu and Osiris in ancient Egypt and the Hindu deity Chandra, who is often worshipped as a fertility god. It might be that moonlit nights gave more possibilities for courtship, hence the Moon being so frequently tied to fertility.

The ancient people also explained cosmic occurrences, such as eclipses, through stories. For example, according to Hindu mythology Rahu is a monster who ate the Sun and the Moon, but was unable to keep them in his gut.

The eclipses represented the occasions when the Sun or Moon was swallowed, but only temporarily. Still, these eclipses were feared as times of disorder, chaos and danger.

In Asia, it is a common belief that the formation of the Moon's surface represents a rabbit or hare – again the symbol of fertility, immortality and rejuvenation. In Sanskrit, the two words 'Moon' and 'hare' are almost identical. Hares or rabbits are also connected to the Moon in Mayan and Aztec cultures.

In European tradition, we often talk about 'the Man in the Moon' and in many stories the man was taken by the Moon as a punishment for some sort of crime or disrespect of the Moon. There are several Estonian stories about a woman going to fetch water late one Saturday night. She is exhausted from the long day and blames the Moon: 'You, lazy one, just watching me from the sky instead of helping to carry the water!' The Moon snatches her as a warning to others about working late at the weekend and showing disrespect. Other stories say that the shadows on the Moon's surface depict a hunter carrying some sort of weapon.

Lunar calendar

Ancient Sumerians were the first to create a calendar based on lunar cycles, so the Sumerian year consisted of 12 cycles. However, this was not exactly in sync with the solar year and the seasons, so an extra month was added every four years.

The Babylonians were the ones to develop Sumerian efforts further and build the system of astrological houses and zodiac signs. They observed that the events on Earth happen in a cyclical pattern, depending on the movement of the Sun, Moon and the planets. Hence, they were the first to do something similar to Moon mapping – tracking the moons' movement across the sky and studying the impact it has on our lives. (We'll learn more about Moon mapping in Part 4.)

Native Americans also used the Moon as their calendar, counting time from one New Moon to the next. Every Full Moon had a specific name, so, for the Native Americans, they became the milestones of seasonal changes and the basis for understanding nature around them – mainly with regard to the behaviour of animals or plants. These names were different across tribes, according to the key sustenance for the particular community. For instance, tribes that mainly ate fish called the Full Moon in August a Sturgeon Moon, whereas communities that ate a plant-based diet might have called it a Green Corn Moon. January Moon was called a Wolf Moon, due to howling wolves at that time of the year. February sees the Snow or Hunger Moon, midsummer is Berry or Thunder Moon, and the most common is probably the Harvest Moon in autumn, nearest to the autumn equinox.

> The word 'Lunatic' – from the Latin *lunaticus*, meaning 'of the Moon' – was used to describe people who were deemed to be mentally ill, crazy or unpredictable. But perhaps they were just people who followed their own flow and lived in alignment with the ever-changing cosmic breath.

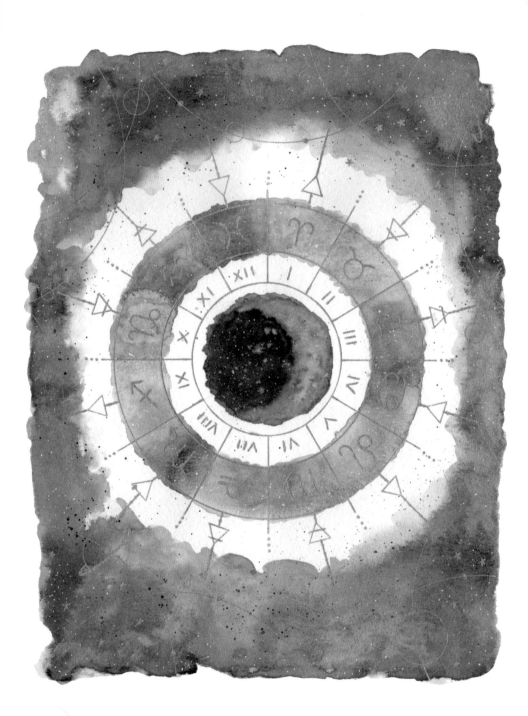

WHO IS MOON?

Also for the Celts, Harvest Moon marked an important milestone in the annual calendar as the time to harvest their crops and organize festivities to celebrate the abundance and blessings of Mother Earth.

Similarly in China, each mid-autumn is the time for a big Moon celebration – Moon Festival, which happens on the 15th day of the eighth lunar month, during the Full Moon. Celebrated in honour of Moon goddess Chang'e, it is the second most important celebration in Chinese tradition, after the Spring Festival, and sees families coming together, admiring the bright lunar light, eating Moon cakes – round pastries with a sweet filling traditionally enjoyed and given as gifts at this time of year – and asking for abundance, harmony and good fortune.

The Crescent Moon became the symbol of Islamic culture in the 15th century. In fact, Ramadan, the month of fasting, starts at the New Moon of the ninth month of the Islamic year and ends with the next New Moon. For Islamic culture, following the lunar calendar is second nature.

In the Christian calendar, the time to celebrate Easter is set by the Moon. Easter Sunday is the first Sunday after the Full Moon that occurs on or just after the Spring Equinox.

I was surprised to find out that in Estonia, which is on the other side of the world from the habitat of Native Americans, February was also called the Month of the Wolf – as it was the wolves' mating season – and when the hay ripened in June, we called it the Month of Milk because the cows that ate the hay would produce a lot of milk. Although the population in Estonia is rather small, different communities still had different names for the Full Moons, as did the Native American tribes.

Many ancient calendars were lunar calendars, based on the phases of the Moon. The survival of the human race depended on their knowledge of nature, the land and the solar, lunar and seasonal cycles, which we have now largely forgotten and replaced with imported supermarket food that is available all year round.

However, the Universe is calling us back. Time has become relative. We're on the brink of understanding that man-made rigid ways of thinking, measuring and categorizing won't work in the future we are trying to manifest. Stress, burnout and various health issues are the direct result of being out of sync with the patterns of nature, being deaf to the whispers of the Universe.

LUNAR GARDENING

In ancient Babylonia and Egypt, our lunar friend was regarded as an invaluable guide for the best times to plant and harvest in order to ensure abundant and healthy crops. The belief behind lunar gardening is that the phases of the Moon impact the movement of hydration, minerals and other substances from one part of the plant to another – in a similar way to how it creates tides.

There are three methods used for lunar gardening: synodic, biodynamic and sidereal. The synodic method is the simplest one and uses the Moon's phases to guide activities. For example, when the Moon is waxing, fluids move upwards, towards the tip of the plant, reaching its peak during the Full Moon, when it's said that the upper parts of the plant are at their juiciest and most full of energy. With the waning Moon we get the opposite effect – the sap flows down towards the roots. Hence, we should harvest vegetables, flowers or herbs with edible blossoms or leaves during the Full Moon. If the roots are what we're after, we should gather them just before the New Moon.

German-born Maria Thun (1922–2012) dedicated her whole life to researching and experimenting with Rudolf Steiner's philosophy of biodynamic gardening. She was especially interested in lunar movement through the zodiac signs and its impact on crops. From her experimentations during the 1950s she divided the days of the month into root days, leaf days, fruit days and flower days, indicating when to show certain plants for the best yield.

It is said that one should never do gardening during the lunar eclipse.

Moon phenomena

The Moon is a dynamic creature. With the growing interest about the Moon and astrology, there are a variety of lunar terms that have reached the mainstream media in recent years. Despite a number of articles and mentions, not many of us really know what these expressions actually mean or where they come from.

Blue Moon

A blue Moon is a second Full Moon that appears within a single calendar month, hence the phrase 'once in a blue Moon'.

Supermoon

The term 'Supermoon' came about in 1979 and is in astronomical terms called a perigean Full Moon – a Full Moon occurring near or at the time when the Moon is at its closest point in its orbit around Earth. Supermoons achieved celebrity status when three occurred in a row at the end of 2016, the November 2016 Full Moon being the closest in 69 years. The next one, which is going to be even closer, will visit us in 2034.

Blood Moon

Traditionally, 'blood Moon' was just another name for Hunter's Moon. It is the second Full Moon after the Autumn Equinox, usually appearing in October and marking the start of the hunting season. As the moonrise is close to sunset in October and November, the Moon can appear with an orange hue.

In recent years, 'Blood Moon' has been used to describe the Moon during total lunar eclipse when she gets a 'blood red' hue. Although the sunlight is blocked by the Earth during the eclipse, the reddish hue comes from the indirect light from the Sun that reaches the Moon's surface via the Earth's atmosphere. The shades can vary from red to orange to gold, depending on the amount of dust particles, water droplets, clouds and mist in the Earth's atmosphere at the time.

WHO IS MOON?

Moon time and feminine power

The length of the female menstrual cycle is on average the same as the time it takes for the Moon to circle our planet. The phases of the Moon can be used as metaphors for the menstrual cycle: Dark and New Moon symbolizing the monthly bleed – clearing, cleansing – and Full Moon the expansion of ovulation – a time of fullness, external focus and high energy – leading again to the release phase.

In days long gone, when we lived without electricity, technology and long-haul travel, the cycles of women were much more in sync and might have actually run in parallel with the lunar phases. In Estonian, my native language, having a period is called 'Kuupuhastus', literally translating as 'Moon cleaning'. Within Native American cultures, even today, having a period is called 'being in one's Moon time'.

In my work with women, I've found that most believe that it's better to have their periods correlating with the inward-time in the lunar cycle – the New Moon. In fact, Full Moon bleeds – also called a 'Red Moon cycle' – were traditionally associated with healers, wisdom keepers and the archetypal Wise Woman as during their 'Moon time' women were believed to be more sensitive than at any other time and,

therefore, more receptive to the lunar energies and cosmic wisdom.

Women on their 'Moon time' are in their most powerful state and going through a natural purification, while being renewed in the process. In Native American cultures it was believed that a woman's power during this time is so strong that it could interfere with sacred ceremonies. By being in a sensitive and intuitive state, women were believed to already be in ceremony during their 'Moon time'. Although nowadays, I certainly would not want to imply that women should keep away from gatherings and celebrations during their 'Moon time', I have personally found that my natural desire is to go inwards and create my own ritual during this time.

Women who are on the Pill or post-menopause, or not having a natural cycle for any other reason, can still use the Moon as a map. Instead of tracking your menstrual cycle, you just track the movement of the Moon from New to Full and back. The Dark and New Moon phase marks the time with similar energy to the 'Moon time', one's period, when we naturally want to draw ourselves inwards and rest. The Full Moon symbolizes ovulation – a time of outward action and expansion, showing your full self to the world.

Moon lodges

Women on their period were considered sacred in countless traditions and cultures. It was believed that the days of their bleed marked an auspicious time, when women had direct access to universal wisdom, invaluable for the longevity of their tribe.

To give women the space and time to rest, rejuvenate and connect to the guides and spirits, and each other, they retreated to Moon lodges or red tents for the duration of their 'Moon time'.

The Moon lodge was the place they could honour their flow, by resting and connecting with their body and its wisdom: sitting in meditation, in silence, or quietly sharing stories with the women around them. Literally, this was the time to be 'in one's body' – in reverence of its innate fertile power.

Being away from the buzz of the village and their daily tasks, women were also in a different relationship with Mother Earth. In a Moon lodge or red tent, they could visualize their roots going down through layers of Earth and offering their nutritious blood back to the soil, clearing everything that was not needed, both physically and energetically.

MUGWORT

Mugwort, or *Artemisia vulgaris*, is connected to the lunar deity Artemis, the maiden hunter. It is also the plant totem of Grandmother Moon and is associated with the Full Moon and summer solstice. Even though it is a sacred ritual plant, you can find it pretty much anywhere, even in urban areas – and it's considered a weed! Although suppressed by concrete, it still has the resilience and perseverance to keep going. Isn't that the symbol of the feminine? Mugwort is great for dream work, strengthening your intuition and working with your subconscious. I like to drink mugwort tea in the evening, or before going to bed, to help me sleep more deeply and to intensify my dreams.

How to make mugwort tea
1 Use 1–2 teaspoons of mugwort leaves per cup of boiling water.
2 Infuse for 10 minutes.
3 Strain before drinking.

Warning: do not use mugwort when you're pregnant or breastfeeding.

Grandmother Moon

In many Native American traditions Grandmother Moon was a symbol of feminine power, connecting women to the cycles of life. As she is the 'weaver of tides' of Mother Earth, she also weaves the cycles of women – like a loving grandmother who helps us to get through life with her ancient wisdom and comforts us in her embrace, allowing us to express our emotions, release and rejuvenate.

Grandmother Moon is the ally of the divine feminine: someone who loves, unconditionally, our ever-changing nature and encourages us to step into our unique truth.

Your lunar birthday

In addition to your annual 'solar' birthday – called the solar return, as the Sun returns to the same place it was at your birth – you also have another one. Your lunar birthday marks the placement of the Moon at your birth and is just as important.

The Moon represents your soul journey – how your path unfolds daily – while the Sun symbolizes your life purpose, your destination, the part of you where you are meant to reach the highest manifestation of maturity. For example, my Sun sign is Virgo, which helps me to understand my purpose:

☽ to serve and heal.

☽ to become self-sufficient and strong in my core and in my own being.

☽ to learn to use Mother Earth as the source of energy, wisdom and power rather than believing that only changing external factors – jobs, relationships or countries – will take me out of my own growing pains.

The way I need to go about my life is informed by my Moon sign – Pisces:

☽ the most sensitive of all the signs, feeling everything and believing in universal love and kindness.

☽ the dreamer and mystic, who needs ample solo time for being and creating.

☽ who needs to be by the water to cleanse.

The more I can honour my Piscean energy in my daily life, and understand my unique needs and challenges, the quicker I will be able to master the highest manifestation of Virgo energy as a practical healer.

Assuming that you already know what your Sun sign is, you can easily find out your birth Moon sign and natal Moon phase via a variety of online resources, by entering the date, time and location of your birth (see Useful Resources, page 157). Once you know these, take a look at the qualities on pages 46–7 and see how they resonate with you.

aries   taurus   gemini

cancer   leo   virgo

libra   scorpio   sagittarius

capricorn   aquarius   pisces

What does your Moon sign say about you?

Moon in Aries – this Moon sign gives you courage and determination to start things. You don't get stuck in situations that won't work for you, but easily change lanes and take the best possible path forward. Use this enthusiasm to share your gifts!

Moon in Taurus – you need stability and security to feel good in the world. Feeling grounded in your body and safe in your space is essential to be the best possible manifestation of your being.

Moon in Gemini – you love communication and information. Connecting with new people and learning something new is essential for you. There's so much to see and explore, so use your curious mind for creating the best version of yourself.

Moon in Cancer – this Moon sign is probably most affected by the movement of the Moon. You want to nurture and be nurtured. Home and safety are important for you as well as the connection with your female ancestry. Honour your emotions.

Moon in Leo – create and express. Be who you want to be. This Moon sign gives you heart-centred courage to love and share. But only to the ones who are worthy. See yourself as royalty and only accept the best.

Moon in Virgo – practical and analytical, you have an amazing ability to see every detail and do the best job possible whatever you put your mind to. No wonder that everyone wants to hire a Virgo. Try to balance perfectionism and work with enough time in nature.

Moon in Libra – you need beauty, harmony and balance within and around you. You also work best in partnerships, so don't go alone. Choose your partners for work and romance wisely, because being in harmony with their energy is what helps you to strive.

Moon in Scorpio – this is a mystical aspect and you probably have emotions, feelings and wisdom flowing like an underground river beneath the surface. You are likely to experience cathartic alchemic situations that will change your life forever, with the purpose to get you to align with your deepest truth.

Moon in Sagittarius – you have a joyful and hopeful outlook and need to spend time learning, exploring and travelling to reach your star. Open your mind and explore all possibilities.

Moon in Capricorn – this Moon sign adds a more structured, serious and hard-working aspect to your life. The way to reach your purpose needs to be practical with feasible step-by-step actions. Keep going!

Moon in Aquarius – this adds an innovative, mind-expanding layer to your being. You need to have time to explore, create and collaborate, knowing that you're touching the edges of what we think is possible. You are the lighthouse of inspiration!

Moon in Pisces – you are more sensitive and emphatic than a majority of the population, so honour it! You are here to love and believe in the best of humanity – the world needs you.

What does your natal Moon phase say about you?

As well as the descriptions of qualities for each Moon sign given above, you may find special familiarity with the sign of your natal Moon.

Born on or around New Moon – similar to Aries Moon sign, you are best in igniting anything new with your impulsive enthusiasm and courage. Life is about you – you are here to shine, just don't forget others on the way.

Born on or around First Quarter Moon – most similar to Aquarius Moon sign, you should be good at building new structures and frameworks, like putting scaffolding up before the builders come. You are inspired by innovative projects and ideas. You act as a bridge between the subjective and the collective.

Born on or around Full Moon – you strive when working in collaboration with others and fulfil your purpose not by going it alone, but by being the connector of past, present and future and the bridge between different disciplines, beliefs and people. You are here to bring ideas into reality. This phase is most similar to Libra Moon sign energy.

Born on or around Last Quarter Moon – if you are born during this phase, you are here to disseminate, analyse, categorize and share information. Just like a journalist who researches a topic in depth and conveys it to the public in a way that is beneficial for our understanding or growth. You have the capacity to see the bigger picture and can break it down into chunks for others to grasp and work with.

PART 3
Your relationship with the Moon

Moving towards cyclical living

In this chapter we will look at the different aspects of the Moon's journey and use an understanding of its different phases to start compiling your personal Moon Diary.

How to track the Moon's journey

Although it's possible to track lunar phases with the naked eye, it's much more difficult to establish which sign the Moon is moving through at any given time.

This is where we can turn to our friend the internet. There are many places where you can check the real-time location of the Moon and its relation with star constellations or zodiac signs – the phase and the sign the Moon is going through at the specific time. I've included a list of websites you could try in the Useful Resources section on page 157.

Mapping the Moon's journey through the sky and star constellations gives us the information to understand the quality of energy each day, the divine timing we have been given by the cosmos.

Knowing this allows me to schedule in extra solo-time for when the Moon is in Pisces (my Moon sign), book a massage when the Moon goes through Taurus, arrange a brainstorming meeting with a team during Aquarius Moon or a healing session when the Moon is aligned with Scorpio.

You'll find detailed descriptions of the energy or quality of each sign in Part 4, but a great deal can be learned from your own observations. Start paying attention to how you feel in the morning when you get up, what you feel called to do, if you are feeling social or would rather hide in the closet, how much physical energy you have, what desires and emotions are coming up and so on – this is how you'll start embodying the ebb and flow of life, the way of the Moon according to your unique experience, and not just by reading this book.

The Moon phases

There are eight major Moon phases that we can observe with the naked eye:

1 **New Moon**
2 Waxing Crescent Moon
3 First Quarter Moon or Half Moon
4 Waxing Gibbous Moon (gibbous means 'convex', 'curving out' or 'extending outwards', which means the Moon is more than half a circle but less than a full circle)
5 **Full Moon**
6 Waning Gibbous Moon
7 Last Quarter Moon or Half Moon
8 Waning Crescent Moon

In this chapter, I describe the physical and energetic implications of five different phases that I personally observe:

1 **New Moon** as the time of quiet creativity (see page 56).
2 **First Quarter Moon** as the time to get active (see page 58).
3 **Full Moon** as the peak of fulfilment, revelation and illumination (see page 61).
4 **Last Quarter Moon** as the time to gather feedback and let go (see page 64).
5 **Dark Moon** While this is not really a 'visible' phase, it marks the time before the New Moon and is energetically significant as the time for rest and letting go (see page 66).

Plus the eclipse phases:

Solar eclipse (see page 68).
Lunar eclipse (see page 70).

We are born
into a world of
natural cycles.
This is the time
to remember the
power they hold.

1. New Moon – quiet creativity

When we can see the first sliver of silver back in the sky after a few nights of darkness, the New Moon is born. The way I always recognized it as a child was that, from where I could see it, it resembled the letter 'J' and, as my grandma said, J is *'juurde'* – 'to add' in Estonian. This meant the Moon starts growing, or waxing, from this point onwards – you can see her growing a little fatter every night until she is Full.

New Moon is like the beginning of the new month. We get up, brush off the debris of the past, take a deep breath and start anew. Think of it like a monthly accounting system. You did your taxes and analysed incomings and outgoings at the end of the previous month, submitted them to the necessary organization and are now ready to focus on the future again, planning and dreaming.

This is a quiet, inward-looking time. Like spring, we slowly wake up from winter slumber and sniff the air for the signs of fresh leaves and buds appearing. We shed the winter darkness and the shadows of the night. Use this time for nurturing your creative plans and ideas. Plant your seeds, write down your intentions. Play. Dream. Go easy.

Similar to taking your time to wake up in the morning, perhaps having a cup of tea in bed, doing some simple movements or breathing...the New Moon welcomes in the new dawn.

This is a good time to strategize, envision where you want to go and allow the answers to come through stillness or nature. Draw mind-maps, journal, be creative, play with ideas and options. Retell your story. Reconnect to your purpose in a new way.

It's also a good time to come together and brainstorm with others; just make sure you choose wisely which energies you are letting near and contain your own power as much as you need. It's perhaps not the best time to let your fresh ideas be criticized by someone who hasn't walked the path. Like protecting your buds emerging from the soil from someone who would mindlessly trample all over them and thereby kill something that could develop into the most beautifully blossoming flower.

2. First Quarter Moon – get active

First Quarter Moon is the half-circle in the sky, halfway between the New and Full Moon, about a week after we saw the first sliver appearing in the dark night. This is where our energy really starts to pick up.

If you'd like to cut the Moon's month in half, one half for pulling inward and reflecting, and the other for outward focus, action and social interactions – then the First Quarter Moon is the turning point. It's the milestone for the energy to turn from inwards to outwards.

It is said that the Quarter Moons (as in First and Last Quarter, about a week each side of the Full Moon) can be somewhat challenging periods as they mark crisis points. We might feel under pressure to take action during the First Quarter. The cosmic energy is pushing us to put our carefully nurtured dreams out there. Like a chick breaking out of the shell, ready to mingle with other chicks and the wide world. Or a caterpillar emerging from the cocoon as a butterfly. This is the time when all the social, outward action starts.

Take action on the intentions seeded during the previous week and kick-start your projects: reach out to potential partners, clients or collaborators, buy a domain and start building your website, apply for a loan. Do a test run or soft launch. Anything that you want to build and grow – align it with the waxing Moon.

YOUR RELATIONSHIP WITH THE MOON

3. Full Moon – fulfilment, revelation, illumination

Full Moons are magical. You don't need to be inclined towards any spiritual path or belief to feel the strange flicker of cosmic touch when you look at the Moon in its fullness. We're reminded that there is so much that is unknown to us, and that our home planet is just a speck in the big Universe.

If you are in the countryside or a forest with no light pollution, you can truly experience a bygone era, when the phase of the Moon played a considerable role in the lives of individuals and the community, informing and guiding their daily activities. Try walking along a dark forest path with Dark Moon – you'd be tripping over every tree root. Full Moon, however, illuminates all the paths in front of us, even the ones we might not have noticed earlier, in the daylight.

This is the task of the Full Moon – illuminating ALL the paths, not just physically, but also metaphorically.

> **During the Full Moon the Sun illuminates the Moon so she can shower us with her full presence, symbolizing the time when we too should let our light shine brightly.**

The Moon is like a big celestial battery, charging our energy to the max, sometimes to the point of overflow. This explains why we might be sleepless during and around the Full Moon and why our emotions may feel heightened. Have you ever noticed that the streets, restaurants and yoga studios are often brimming with people around the time of a Full Moon? An increase in crime rates, road accidents, acts of vandalism and calls made to emergency centres during the time of the Full Moon has also been noted by members of the police force and nurses. While staff in psychiatric hospitals have observed that some patients are more agitated during the Full Moon.

Science, astrology and spirituality got divorced a couple of thousand years ago, so they are not friendly bedfellows in the eyes of many, but my opinion is that we all need to make up our own minds and follow our intuition when it comes to looking at quantitative studies or our own physical and energetical state.

Full Moon intensifies everything: sensations, emotions and situations reach their peak in order to show us what needs to go, or be transformed. It shows the consequences of the steps we have taken and the ones available to us going forwards. It helps us to discern what's in alignment and what's not, change focus and redirect our energy. The Full Moon is called a 'climax in a process' – the peak of the outward manifestation of our inner work.

Need to make a decision? Work something out? This is the time to sit with it and literally receive the solutions sent to you by the Moon. It's the junction point that lights everything up and asks us to make the choice.

The fullness of the Moon expands our energy and social skills, so if you need to make a pitch or presentation at work, or organize a party or a social event, this would be the best time to do so. Also, launching anything that requires public feedback – such as a website or shop – would be best done on or around the Full Moon.

It is possible that a lot of revealing and transformative information wants to come through during these days, which can be quite intense – so how to keep flowing with the beautiful lunar waters?

We can balance it ourselves. When the energy pulls us higher, we need to keep ourselves grounded. If Full Moon energy feels overwhelming, I strongly recommend making sure that you reduce or cut out intake of any stimulants for a few days – this could be alcohol, caffeine or sugar, for instance, or too much screen time, loud music and so on – to let your nervous system settle and deal with the lunar transition.

Our bodies and energy levels can easily be over-stimulated and over-energized, so if you add all the extras, it can leave you ungrounded, rather than able to use the powerful cleaning and healing energy. To receive the gifts of Full Moon we need to slow down, even though the energy tempts us to speed up and go full force. If we ground ourselves and channel the (sometimes boiling-over) energy into movement, creation, healthy expression and deep connection, we come out as winners.

Full Moon is like a loud bang to wake us all up, bringing whatever might be hidden or unseen – emotions, challenges or opportunities – to the surface, making them visible. Arguments, upheavals, emotional outbursts, tears, physical and mental exhaustion, confusion...like they say, feel it to heal it.

It's not only human beings that are affected by the Full Moon. The light of the Moon is the trigger for the spawning season for coral on the Great Barrier Reef in Australia. A report by Dr Frank Brown, an expert on animal behaviour, says that hamsters spin in their wheels more aggressively during the Full Moon. A study has also found a correlation between the rise of injuries in cats and dogs and the Moon's fullness – I definitely notice our dog Lune becoming more restless and irritated during Full Moon. Have you noticed your pets' behaviour change? It might seem odd, but having a Moon Diary for your pet (together with your own) is quite a good way to see if and how the ebb and flow of cosmic energies affect your beloved furball. (More on Moon Journalling on page 72.)

Beliefs traced back to ancient Greece and Rome say that the Moon, particularly Full Moon, changes the level of moisture in the brain, consequently bringing about altered states of mind. It was also believed that different phases of the Moon changed the composition of the fluids in the body, such as blood, phlegm and bile.

4. Last Quarter Moon – gather feedback

Last Quarter Moon is the next milestone, where the tide turns inwards again. Our unease with this phase – or any time that we are made to 'slow down' – is a powerful example of how the current growth-based economy is misaligned with the natural world and with the cosmic breath: we are living in an economy where the goal is continuous growth, where production of excess is the marker for success, and where we measure progress with numbers and financial assets.

In our natural state we should only be 'active' and 'growing' half of the time. By observing the seasons, the cycle of night (sleep) and day (action) or inhale (inspiration) and exhale (clear), we understand that the world is not just one long summer day with a permanent inhale. Life on Earth would not last very long if that were the case. It's essential that we honour and protect the slowness, reflection, unplugging from the information society around us and retracing our steps in order to restart with more clarity and direction.

The waning Moon phase is a good time to identify distractions and energy leaks. How can you create the life that supports your creative gifts – manifest the money you need, the space you need and the people who support your growth?

Any sort of analytical work is good here: like harvest time when you have a pile of fruits and you are asked to separate them into the ones that are good to eat (enjoy!) and the ones that need to be composted, cleaned and transformed into a different state (jam?) or given to somebody else who'd appreciate them (dog?).

Look at your numbers, look at your data, look at your feelings. See what works and what doesn't and take no nonsense. This is the time when our diplomatic buffer is thinnest, so we have an eye for detail and we are inclined to call out anything that is not right or in alignment.

This is your time to clear and clean – take care of everything you want to diminish, eliminate or let go of in your life. The body also wants to rest: support it with intermittent fasting or eating leaner and cleaner as well as being more gentle with yourself while you turn inwards for reflection. Clean up your desk/your drawers/your home and tick off everything that has occupied your to-do list for too long – they will be blocking your energetical flow otherwise.

5. Dark Moon – let go fully

When the Moon is not reflecting any sunlight we're in the Dark Moon phase. You're welcome to float into the unknown and enjoy the void. I usually count back three days from the New Moon and mark it in my calendar as the time for a mini-holiday. Isn't it welcome news that the Moon grants us a 'mo(o)nthly' break? Get comfortable with it and learn the power of rest. This is the deep winter of the lunar month.

Work wise – sit still and receive valuable insights and inspiration from your unconscious. Detox the old in your space and in your body. The more space you make during this time, the more space there is for new when the next cycle starts.

Stop being afraid of slowing down and enjoying the 'be, heal and receive' mode. Most of us are familiar with the deep-seated fear of jumping from the conveyor belt of busyness. We are stuck in the mud of 'constant productivity' that has been thickened over the last centuries. We are afraid of the 'death' of our ego when we take our foot off the gas. The world won't stop. Instead, this is when you are welcomed into your true life.

Each of us has been given individual tools and the capacity to remain calm among the chaos, among all our chores, to-do lists and responsibilities. Learn to use them.

For me the Dark Moon is the time of detox, cleaning and clearing – physically and emotionally as well as the environments around us. Like a small finale for making space for the next act. I naturally eat less food when the Moon is waning and make sure it's as clean, fresh and simple as possible. The body is in the state of letting go, so help it along. Schedule in time for rest and recuperation during the Dark and New Moon. Getting enough sleep becomes especially important now.

Solar eclipses – the reset

The eclipse is a time of rebirth. The Sun's rays are temporarily blocked from mirroring back to us, creating a sort of void or cosmic chaos where everything becomes possible. Your pack of cards can be reshuffled.

Solar eclipse occurs a few times a year, when the Sun, Moon and Earth are aligned. Total solar eclipse happens when the Moon blocks the Sun completely, making it possible for the dark shadow of the Moon to obscure the light of the Sun, leaving us only with the halo – an aura of plasma that surrounds the Sun, called solar corona, most easily seen during a solar eclipse.

The effect of solar eclipse on animals has been widely studied and it appears that animals and birds that are active during the day return to their homes or to nocturnal activity, while night-time beings think they've overslept. It has been found that, during an eclipse, some spider species start breaking down their webs, which typically happens at the end of the day. Once the eclipse is over, they begin rebuilding them. Solar eclipse appears to make animals confused as the usual day–night cycle is disrupted in a rather unusual manner.

As humans are part of the animal world, living and breathing with the cosmic forces in and around us, it's clear that the occasion should similarly stir something in us. There are many myths and stories that explain the eclipse in different ways, but the underlying theme is one of grand 'reset' with a definite change in the vibration and frequency of energies on Earth.

If we think about it, during the solar eclipse, the Moon – HOW we do things, the process that unfolds our main purpose – aligns with and obscures the Sun – our purpose – thereby opening a portal for redirecting our life story and redefining what our real purpose here on Earth is.

For me, the eclipse season feels like a David Lynch film: we're abducted to a different realm where the senses and emotions are heightened, colours, sounds, spaces get an otherworldly, dream-like vibration, everything seems a bit more intense. As we are taken out from our everyday reality and our minds and energies are stirred up, as in a cosmic tornado, we emerge slightly dishevelled with a different perspective on our past or current beliefs and activities. The eclipse gives us the opportunity to rebuild our life with new bricks.

Native American shamans believed that a solar eclipse is a powerful time of healing. As the Sun, Moon and Earth align, it realigns us by resetting our purpose on an individual level, dissolving differences and creating a coming together as One on a collective level.

As solar eclipse occurs together with the New Moon, it marks the time of the new cycle in an amplified form, like giving us a supercharged New Moon. We can feel tired and lethargic during a solar eclipse, but this is usually the means for the Universal energies to get us to slow down and go into stillness or dream-state, which is where the real magic happens and the seed can be sown.

Lunar eclipses – final letting go

Eclipses, both solar and lunar, always happen in groups of two or three – called an eclipse season. They mark a time of powerful and unexpected stir-ups. Eclipses can bring up situations that accelerate our growth – such as new opportunities that appear to come out of the blue, or obstacles that get thrown into our path. In other times, the eclipses make visible the old patterns of behaviour that we need to face up to.

For a total lunar eclipse to occur, the Sun, Moon and Earth need to be in one straight line, while the Moon passes through the darkest part of the Earth's shadow – the umbra – and becomes coloured with a deep, dark-red hue (see 'blood Moon' on page 38). Basically, the Earth is sandwiched between the Sun and the Moon, leaving the Moon in shadow. We can also have a partial lunar eclipse, or penumbral lunar eclipse, when the Moon travels only through the outer part of the Earth's shadow.

Lunar eclipse has a slightly different alignment to solar eclipse and happens during the Full Moon. As the Earth moves between the Sun and the Moon, it cuts the energy flow between these two celestial bodies. We can look at this as a time of possibility – to get out of an old cycle, to re-shuffle our pack of cards. It can feel like the door to our unconscious suddenly opens and we are able to go in to clean up the house, or like someone dives into our subconscious to bring old emotions and attachments deep from our psyche to the surface to be processed and released – to be immersed in healing light.

Any childhood, ancestral or family trauma can easily come up with the eclipse, so be extra gentle with yourself during this time and stay connected to your body and Mother Earth. The best ways to keep rooted and grounded in your body are through breath and movement. It really doesn't need to be anything complicated, even something as simple as taking five mindful and long inhales and exhales, or dancing through a song and allowing your body to move freely, will be a great help. Shaking your body is also a good way to release tension and get back to the present moment. I always encourage reaching out for guidance and support – it can be from a therapist, a reiki master, an energy healer or even a good friend whom you completely trust. We are not meant to grow on our own.

For me, solar eclipse feels like a shift with more 'external' impact and a sense of looking into the future, while lunar eclipse is about healing our past and has an 'internal' vibe. At the end of the day, the goal of eclipses is to bring us into alignment with our cosmic path, life purpose and Universal breath.

How to start your Moon Diary

Now you know how to recognize the key Moon phases, you can start keeping your Moon Diary – this will be like your new calendar and will evolve over time.

Focus on tracking the phases in the beginning and then add the signs when you feel ready. We will go on to explore the signs in Part 4, with specific rituals or exercises you might wish to try for the New Moon or Full Moon in a particular sign.

There are so many different ways to keep a Moon Diary, so find a way that works for you – whether it's typing things up on your phone or computer, writing or drawing on paper, or using a voice memo on your phone. I still find using a paper diary or blank notebook works best for me.

Step 1 – Tracking

Before you check in with the Moon, check in with yourself. The idea is to learn to read YOUR energy flow. Trust your inner authority first. Your personal Moon Diary is not a random newspaper horoscope column, but a profound guide for deeper self-awareness. A process that will help you to tune into the cyclical spiral of life.

Write down a few words or sentences in the morning or evening. This will help you to check in and see where you are. Note your observations and insights from the past 24 hours – how did you feel? How did you relate to others? Did you crave some solo-time or have boundless energy for social interactions? Did you feel creative or analytical, motivated or stuck? How were your physical energy and confidence levels? Note down anything that comes up.

If you are someone who has tried and failed to commit to daily journalling in the past, start a one-sentence journal. It's hard to find excuses to not make reflection a daily habit with this one.

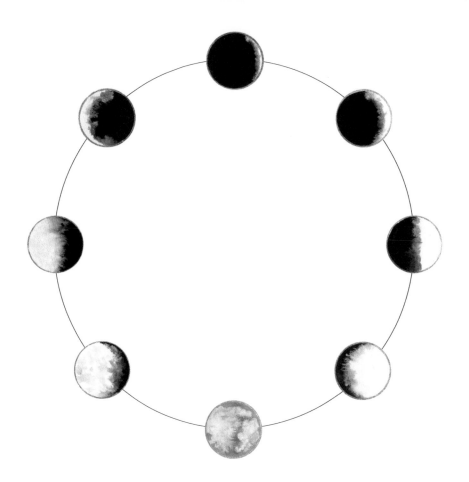

After you've jotted down whatever came up – thereby deepening your self-awareness – check in with the cosmic energies. Note down where the Moon is on its journey – for instance, New Moon, waxing Moon, Full Moon, waning Moon. (You can check what phase the Moon is in using one of the websites I recommend in the Useful Resources section on page 157.) When it's New Moon, mark this as Day 1 in your Diary. The Moon starts waxing, hitting First Quarter (half-circle) in about a week; Full Moon should appear around a week from there, Day 14 – marking the peak of the journey – and then starts waning again, going into darkness before the new cycle begins.

The next step is to add an extra layer by checking what sign the Moon is in, as well as the phase. (Part 4 details the energetic qualities when the Moon is in each of the 12 signs.) You could start by paying particular attention to when the Moon is in your own lunar sign. For example, my lunar sign is Pisces and when the Moon is in Pisces, I generally wake up feeling more raw and sensitive than on other days, and feel nourished when I allow myself time to just be and create (the best time for writing this book was when the Moon was with Pisces).

Tracking things in your Moon Diary in this way will then help you to start planning your own calendar according to your flow which will, in turn, influence how you live your daily life and schedule commitments.

Step 2 – Planning
As you sense the patterns, you can begin to plan particular projects or activities with the Moon in mind. You can look ahead and decide the best time to give that presentation at work or when to book in a quiet weekend for some rest and time in nature, depending on which phase the Moon is in.

Need to do creative work? Schedule it for when the Moon is in Leo, Aquarius or Pisces.

Need to do some number crunching? Choose Capricorn or Virgo, practical earthy signs.

Planning a spa day? Book it in when the Moon is in its watery home, the sign of Cancer.

Step 3 – What you'll gain from keeping a Moon Diary

Once you know when you're most creative, most confident, most social or most analytical, you start trusting your own body and your inner authority over anything external.

- ☽ With setting your own pace, stress becomes a thing of the past.
- ☽ While you are radically slowing down – resting more, playing more – you are actually becoming more creative, productive and successful.
- ☽ You will respect yourself (your time and energy), which makes others respect you (and your work).
- ☽ You'll be able to harness the skills of your colleagues and team by understanding the energy available at any given time.
- ☽ You stop beating yourself up for changes of mood or physical energy and go with the flow.

As you become more familiar with your own relationship with the Moon and her cycles, cherish your place in the vast web of cosmic consciousness.

Are you ready?

Get your diary (or use the template on the following page) and make a start by writing down your ONE sentence for today.

Moon Diary

Today I feel ....
....
....

Phase ....
....
....

Sign ....
....
....

Today I feel ....
....
....

Phase ....
....
....

Sign ....
....
....

Moon Diary

Today I feel ....
....
....

Phase ....
....
....

Sign ....
....
....

Today I feel ....
....
....

Phase ....
....
....

Sign ....
....
....

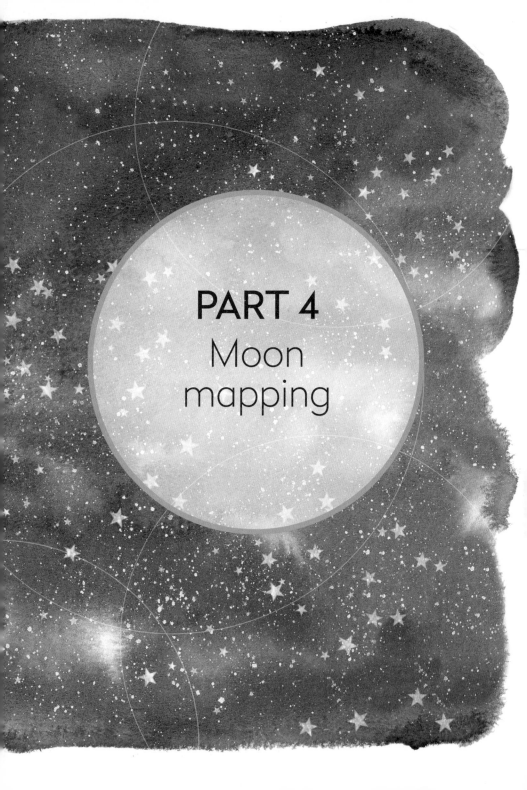

PART 4
Moon mapping

A blueprint for cyclical living and creating

The Moon completes its journey in approximately 29.5 days, meaning that we have either a New or Full Moon every 14 days or so. Also, it changes signs every 2–3 days.

On the following pages I will go through each of the 12 signs individually – giving a general description, questions to explore, how to benefit from the energy of the sign and the best tools to use for each sign. In your Moon Diary (see page 72), you can make your own notes relating to different phases and/or signs of the Moon, as your relationship with *la Luna* is meant to be interactive, practical and personal.

He or she?

It's important to explain my use of he and she/his and hers/masculine and feminine. While this book is for every one of you, regardless of how you identify yourself within the gender continuum, I've included the categories to note whether the particular period of the Moon's journey is governed by masculine energy – extroverted, action-oriented, giving – or feminine energy – receiving, nourishing, flowing, creative. These are just qualities, which are intertwined in all of us, regardless of gender identity – like yin and yang. Also, I might refer to the Moon as she/her every once in a while because of its association with the feminine in many cultures, for example Grandmother Moon (page 43).

Moon energy

For each sign, I have added the elements – earth, air, fire and water – that represent the energies we work with. I hope these categories help with further decrypting the energies available.

EARTH
Fixed, grounded and practical.

AIR
Inspirational, communicative and thought-based.

FIRE
Igniting, passionate, impulsive and transformative.

WATER
Flow-y, cleansing, regenerating and emotional.

Moon in Aries

Aries, being the first sign of the zodiac, is like New Year's Day. He enters with fireworks that won't go unnoticed. Aries is often called the 'baby' among the star constellations. Governed by the warrior planet Mars, he makes sure he gets what he wants, and gets it now – like a four-year-old in a toy store. And if things don't go his way, expect nothing less than a tantrum. Arguments? Hot-headed discussions? Bursts of anger or even rage? This is one aspect of Mars/Aries energy.

On the flip side of the coin, these days are like fuel to fire for those who are not afraid to break the mould and stand in the forefront of big changes. Find your courage. Aries energy is here to make you more assertive and confident – moving away from the over-nurturing, over-mothering archetype to the one that gives clear directions and allows the masculine energy within to take action. Be the power rather than react to it.

Wake up earlier and welcome the Sun when it rises. Run wild and free, and know that everything is possible. Move countries, change jobs and strategies, initiate or leave partnerships – anything to get you closer to YOUR way. Become the fearless and fierce (but compassionate) queen or king of your territory – whether it's personal or business.

This is the energy that invites us to take up space in this world. Breathe in and fill your lungs with inspiration. This is about you, YOU in capital letters.

Invitation – what to explore?
☽ Are you struggling with low self-worth?
☽ Are you procrastinating – perhaps stuck in a relationship or situation at home or work you are not happy with?
☽ Are you repressing your emotions and feelings and consequently feeling resentful?
☽ Are you holding yourself back, paralysed in unconscious fear that stops you from taking the leap into the unknown?
☽ If you'd known you are fully supported, what would you do?

Moon Energy – masculine fire
Aries is associated with the masculine energy. The fire. The warrior. The initiator. These are the qualities so familiar to those of us who have souls like bubbling volcanos. Who want to be rebels, the fire-starters, the trailblazers, make a difference in the world. However, we need to learn to use this fiery energy in a mature way, with certain caution and to not burn ourselves or others in the process.

How to benefit – personal responsibility

If you want to harness the energy of Aries Moon, start by exploring your relationship with personal responsibility. You are the one in charge of your inner landscape, your beliefs, words and actions. You are the one who can initiate change. As Aries is the fiery 'baby', it can fuel heated discussions, arguments and fights with others when we're feeling wounded, hurt or just not completely happy with our lot. However, we need to understand that how we approach and respond to situations is entirely our choice and not down to an external circumstance.

Tools

Breath work is a great tool for breaking through any blocks – like fear or repressed anger – and transitioning into calmness and courage. Although a 45-minute transformational breathwork session might be intense and fiery, the beauty of breath is that it's a portable tool – even taking five deep breaths in through the nose and out through the mouth takes you to a much calmer realm.

Smudging or burning incense – Aries is fire, but also related to air. One of my favourite tools that unites these two elements is burning sacred plants like sage (*Salvia officinalis*) or palo santo (*Bursera graveolens*) to clean my own energy and space. During the day, it's nice to burn incense wherever you work, watching the smoke go up and reflecting on the explorative 'Invitation – what to explore' questions on page 83.

Native to South America, palo santo ('holy wood') was used by ancient civilizations like the Incas and it's available in many places on- and off-line. However, be mindful about the provenance of it as it's an endangered plant – make sure you get it from an ethical supplier or choose another plant that is sourced closer to home.

New Moon in Aries

The first Moon after the Spring Equinox in the northern hemisphere and the Autumn Equinox in the southern hemisphere marks the beginning of the new cycle, a start of the 'Moon year', if you like. Much more than just gentle intention setting, this kicks everything off with fireworks and spontaneity, facilitating the movement through the spiritual birth canal, ready for a change of scene and action, and for a big inhale. This is a powerful time to 'burn' our old selves in the cleansing fire of Aries and collectively 'burn' old-world beliefs and expectations like a wildfire, to make space for fresh shoots.

Modern ritual – schedule a chat

Is there something that nags at you and feels like a block in your heart and/ or throat, asking to be released and articulated? Use the energy of the New Moon in Aries to give you the necessary assertiveness for initiating the conversation. You can do this with your partner, lover, friend, client, colleague or anyone else, but start with someone who is generally understanding and not the most aggressive type (as the Aries Moon can amplify any hot temper). Practise your courage muscle. Practise being you, fully and truly.

The best way to get your point across is to first make sure it's a good time to talk – probably not during a dinner or party, or very early or late in the day. First rule – don't blame, but communicate calmly how the situation makes you feel, what you need to happen and, if these needs are met, what the benefits would be – for yourself and others (in other words, why it is important for you). Don't try this during Aries Full Moon, as staying calm might prove tricky.

Full Moon in Aries

Like a fireball of energy, this Moon gives us the power to move forwards with a full tank (which can as easily overfill) and determination to change our lives for the better. Luckily the Full Moon always helps us to see the areas where changes are most needed. If you know more about astrology or have your birth chart, you can always check which 'house' or area of life this Full Moon illuminates, as this is the area where the transitions happen.

Having said that, Aries Full Moon is not the time to make rapid, impulsive changes. It can be a bit too harsh and brutal, making you take action too soon. Remember that everything has a divine timing and it's important to check in with your body, to see if it's really the right time for anything big. Radical shifts can feel freeing – however, when it comes to your daily life, make sure you don't accidentally throw out the good while throwing out the bad.

Modern ritual – enjoy the view

We need to be in a constant balancing act. The more fire is in the air, the more calmness and spaciousness we need. It seems counter-intuitive, but during Aries Full Moon, learning the power of rest and relaxing your mind is essential.

A beautiful ritual for this time is to choose a rooftop, park or hilltop where you can enjoy a view over the city or landscape. This reminds us that there is so much more in this world. Regain the perspective and clarity that are so needed at this time.

Find your
courage

MOON MAPPING

Moon in Taurus

Taurus is steadfast and needs her hooves strongly grounded into the Earth. These Moon days are about igniting your bull-like stability and self-sufficiency. Return to your body wisdom, root down to Mother Earth and don't let every gust of wind affect your unique experience.

Taurus focuses on the physical plane, the matter, the ability to feel sensations and pleasure through our bodies and see beauty in exquisite objects and the environments around us. Loving sex, pleasure and money does not make us dirty, greedy human beings. This is the time to change these beliefs and past conditioning. The truth is that our true power awakens only when we welcome and inhabit all the parts of our body, understand our body's magnificent power and beauty, and allow pleasure to be our guide. When we love ourselves and open our portals to receive all the goodness there is, we can truly care for and love others.

This is why the flip side of Taurus energy is one of rage, and we might easily destroy everything around us. Our wounds, insecurities and past hurts can stab us like the sharp sticks of the *banderilleros*, making us race around the ring like a raging bull at a bullfight, not knowing where to go or what to do with the emotions we are experiencing.

These days remind us that our bodies are temples of wild power – beauty and pain, strength and intuition. The way we can go from being the angry bull in the ring to the grounded bull munching fresh grass is through deep self-love, pleasure and connection to Mother Earth. When the Moon journeys through Taurus, there is no need for sudden activity. Surrender to the call to slow down and just *be* with Earth. This is the way for Taurus to invite us back into our bodies.

Invitation – what to explore?
- How often do you feel true pleasure?
- Do you feel guilty about pursuing pleasure?
- Where are the energy blockages and places of tension located in your body?
- Can you discern how your body says Yes to some things and No to others?
- What does your body need?

Moon energy – feminine Earth
Taurus is ruled by Venus, the planet of the feminine. Feminine is always about receiving and creating – through human bodies and the Earth. This energy has incredible healing ability as it's the area that many of us are disconnected from. The feminine and the Earth have been robbed of their power for a long time, so Taurus Moon can bring out collective sadness, anxiety, anger or grief, especially if we're not allowing ourselves

to slow down and connect to the incredible source and soak in the beauty at that time.

How to benefit – grounded pleasure

Feeling grounded and safe makes us radiant. This is your monthly invitation to explore Taurus energy and nurture yourself. Feel the magnificent abundance of nature around you. If we only keep climbing higher in spiritual realms, without balancing it with the connection to the physical, manifesting what we truly desire is much harder.

Make joy and pleasure (not just short-lived, but real embodied pleasure that is simultaneously physical, emotional AND spiritual) your guide post, your lighthouse, showing the way and giving a direction to your actions. With every situation, every client, every decision you need to make, you can ask your body and listen to its answer. If it is saying Yes with a deep feeling of pleasure, you'll know how to proceed.

The belief that success means struggle and suffering is the biggest lie and we need to delete it from our systems. Once you let pleasure be your guide and start honouring yourself fully, demanding equal exchange, and never compromising with second best, you'll open the portal for abundance (money, opportunities and connections, job offers, new friends and lovers – anything that you wish for) to flow in.

Tools

Taurus invites us to explore and celebrate our physical form, reconnect with our own bodies, the bodies of others, plants and the Earth. Give yourself time to connect with nature in whichever way and receive her messages and insights.

My favourite tools for Taurus Moon days are **bodywork** of any kind. You could book yourself a massage or try abhyanga, which is Ayurvedic self-massage with warm oils. Also try slow walks in nature, forest-bathing, and enjoying a pot of tea in silence, or perhaps going to a tea ceremony. All of these connect us with the Earth on a deeper, more embodied level.

New Moon in Taurus

New Moon in Taurus has a beautiful grounding energy, focusing on physical and sexual power. We need to allow rest, play and pleasure to return as integral aspects of being alive and a key part of our everyday reality, not just a naughty escape. The New Moon is a gentle invitation to start using these tools as the link between our bodies, nature and our purpose on Earth.

By becoming acquainted with, and self-sufficient about, your own sexual energy you can radically change your life and business. And this is much more than 'having sex' – in fact, the energy we are talking about should be used alone, with reverence and tenderness. I invite you to move with slowness and patience, be mindful where you are and be gentle with yourself. Many of us have deep wounds connected to our sexuality so, if that's the case for you, find someone to guide you on the way, who can help you learn the tools that take you towards pleasure.

Modern ritual – say hello to your pleasure

Finding deep nourishing pleasure is easier than you think. You can find it everywhere, any time. To start getting acquainted with the range of pleasure available to you, try the following practice: lie down somewhere cosy, where you won't be disturbed, and start with slow inhales and exhales to relax your body.

Now, focus on your skin. You can touch it, letting your fingertips run over your arms and legs, or neck and stomach. Breathe in and imagine the surface of your skin is in love with the air surrounding it, tingling with joy. Breathe out and relax. Do it for few minutes and see how you feel – did you find a spark of pleasure in just being in your skin?

Full Moon in Taurus

The bull of the sky is patient and practical until the ground underneath starts shattering. If you don't allow yourself to slow down when needed, to just be and recharge your batteries when you feel depleted, you may notice the power of rage, which can approach as a hurricane. Full Moon can amplify anything that goes on underneath the surface, including wounds that are pulled together with fragile plasters and not fully healed.

There are going to be countless insights and 'aha' moments at your disposal if you connect with the Taurus Full Moon energy through your physical plane. Keep listening to your intuition and body wisdom to clear the cobwebs physically, energetically and emotionally. Intuition shows us the places that we are losing or giving away our strength and grounding.

Modern ritual – embrace *shinrin-yoku* or forest-bathing

How can we create stability for, and around, ourselves to build the resilience to face the storms? The practice of *shinrin-yoku* (the Japanese art of forest-bathing) is perfect for Taurus Full Moon and is said to boost the immune system, lower blood pressure and aid sleep. Leave your phone and camera at home.

The only task you have during your 'green bath' is to be present and enjoy the sensory experiences: birdsong, the smell of moss beneath you, the visual patterns of the trees and leaves or the play of sunlight and shadows. You can just sit in one place or walk mindfully and with no need to get anywhere – just follow your intuition. Proceed with Taurus-like slowness, savouring the journey.

One UK study, carried out by King's College London and published in January 2018, found that exposure to trees, the sky and birdsong in cities improved mental wellbeing. The benefits were found to be still evident several hours after the exposure.

Feel your roots going deep down into Earth and being one with the Earth.

If going to a forest is not possible, just take your shoes off in any spot that has a patch of grass.

Ask for more
pleasure

Moon in Gemini

Gemini is a changeable creature. The (social) butterfly. The time traveller. The one with the 'portfolio career'. He diversifies your story and wants you to live many lives in one lifetime, gaining experience by learning, connecting, communicating and diving into a variety of projects. If you feel stuck in a rut, this energy is your best ally to push you to try something else, to choose a new path. And Gemini's change is not incremental, but radical.

This is the portal the Universe gave us for flipping the switch and changing our life in an instant. Yes, we can do that now. What we think our life is, what we are able to manifest, the stories we tell ourselves about who we are, limitations we set, what we think is possible – these are all works of the mind. Gemini asks us to look at different facets of ourselves. To turn the diamond and look at the beauty hidden on the other side. There is no need to hold onto our old 'identity'. In Gemini, life happens now and in the future. You CAN be everything you want and more.

When it comes to work, this is the time to make things happen. Take action. People are interested in what you do and it will be easier to get your point across – do it with clarity and conciseness as your attention span can also be pretty low on these days. Make it interesting, use your passion and enthusiasm to fuel your projects. Any sort of communication – writing, learning, networking, brainstorming – is ideal and you can get a lot done if you're able to focus and ground. The flip side of Gemini energy is the tendency to become a bit fried, frantic and frazzled with all the information and excitement floating around.

Invitation – what to explore?
Which new skill would you like to learn or explore – maybe reiki, coding or creative writing?

- ☽ Do you feel there's an area of your life in need of radical change – work, your relationship, wellbeing, home?
- ☽ What is getting in the way of you changing your story?
- ☽ How could you inject more childlike curiosity into your life?
- ☽ Who are the five people you could get in touch with who could help you on your journey?

Moon energy – masculine air
Gemini is playful, creative, curious and interested. Hungry for life and eager to taste all its aspects. One with the need to explore many paths simultaneously, being guided by what feels most fun. It's the time of doing, opening the portals of manifestation. Go for your dreams.

This doesn't have to be a serious pursuit, but a playful and lighthearted one. The air element is reminding us to cleanse ourselves of old stories, lighten up, let our hair down and enjoy the ride. Everything is possible.

How to benefit – change the way you talk (including to yourself)

If we change the way we think about ourselves (self-worth!) and how we talk to others, EVERYTHING changes.

We have a choice about how to approach each emotion and situation.

Do you beat yourself up over everything? Gemini is closely linked to our thoughts and words – how we think and talk to ourselves and others is closely linked to what we can manifest externally. When you integrate kindness and playfulness into your speech, life will lighten up, negative self-talk will diminish and, most importantly, you will attract support from others.

Tools

As this is a great time for exploring anything new – places, people, ideas – I find the best tools are online platforms for **wellbeing events** happening near where I live. Some of my go-tos are listed in the References list on page 157.

Subscribe to **podcasts** – when I added listening to podcasts to my grocery trips, I could kill two birds with one stone, and multitasking is something that Gemini loves. My favourites are in the References list on page 157.

New Moon in Gemini

What are the topics that interest you? Gemini New Moon is all about planting the seed of intention of what your future self could look like and asking the Universe for the right teachers, courses and guides to come your way.

Governed by planet Mercury (representing the mind), the New Moon asks you to look inwards and allow the mind to be a loyal servant to your heart, intuition and body wisdom. Don't get lost in the storms of your thoughts and waves of information, but develop discipline in setting boundaries to what you receive and transmit and how. Build intention into everything.

Modern ritual – embrace digital boundaries

It's easy to get lost in all the information available to us and lose ourselves on the way. How about learning to enjoy offline activities – perhaps reading a book, exploring a new neighbourhood, journalling – again? Around this New Moon, limit screen use to 4–5 hours each day and have a day completely offline each week. Also observe what you are using the time online for – are you spending it wisely? Is the information you are disseminating or gathering benefitting your future self?

Full Moon in Gemini

While in continuous hustle mode, we may fail to notice that we're already in the place where we're safe and taken care of. What's the worst that can happen if you let your hair down and play? The more you trust that you will be held, the more the Universe will be able to show you that you are.

This Full Moon is your invitation back to lightheartedness. The opened portal invites you to bask in magic, explore your fantasies and dance like all your dreams have come true. Try it out and see how it makes you feel.

Modern ritual – change right NOW

Live your dreams. Dive into the life you want to live – now! If you're thinking about changes you want to make, having visions about your 'future life' – what are you waiting for? Gemini can jump from one way of life to another like changing outfits.

Want to boost your looks? Never leave the house without red lipstick and a dress or a tailored suit – starting now. Want an LA-lifestyle? Take vitamin D baths in the Sun, stock up on superfood powders and get blending – starting now. Want to change your career? It's easy to set up or change a website these days. You can do it now. Get yourself out there.

Gemini pushes you out of your comfort zone. Once you make the leap, you'll find that synchronicities await. We're all in this together.

You can be _can_ be everything you want

Moon in Cancer

Cancer Moon energy is like a huge eternal mother wrapping her arms around you, comforting and compassionate, bringing whispers of maturity and wisdom. It renews our relationship with the feminine/mother – Mother Earth as the land beneath us, our biological mother or the one who raised us, collective feminine, our ancestral lineage of women, and it also asks how we mother ourselves and others.

We can heal the wounds and trauma of our ancestors and motherland only by processing our own wounds in solitude, by healing ourselves and then letting our nourished light ripple above and below, to all directions, including the past and the future. This is the time to be around water; let the the water soothe and heal you and ask for guidance and healing from women past and present as well as from the land beneath you.

Emotions can be high at this time and we might feel more 'watery' and sensitive. Don't hold onto – or get stuck in – situations, places, things, people, thoughts and emotions, but flow through gently, trusting that you are guided to where you need to go. Like a river flowing between the banks, find security in surrendering to the path set in front of you – the river doesn't question or doubt, it doesn't beat itself up for not flowing faster or regret its path. We have so much to learn from nature and the elements.

Invitation – what to explore?
- What makes you feel safe?
- Where is your sanctuary, your place of peace and calm? Is it an inner state, certain environment or specific place?
- Do you know your female lineage – the stories of your mother and grandmothers?
- How often do you gather strength from the land and her waters?
- Do you have sufficient solo-time filled with self-care?

Moon energy – feminine water
This is a beautiful time of great healing and receiving. Like a river that knows that it is shaped and guided by the shores holding it, gently dancing with the river banks that mould its path. Always moving forward. Exploring. Trusting its track.

If you feel wired and overstimulated (flowing too fast), it might be a sign that you need to simplify your life. Let the debris fall away, so that your base is clear, and just float. Only with proper nourishment, can we Think Big and manifest everything that we dream of.

How to benefit – ask your guides

What's special about this time is that you will be surrounded by a support team of cheerleaders. These are the women of the past, the souls from your lineage, the sisters around you. You can let go, release and know that you're being held, loved and supported unconditionally and like never before. Be willing to trust it.

Call your mother or grandmother, or perhaps an aunt. Ask for guidance from your spirit team.

Tools

This is the best time to spend time by **water**. I find moving water especially beneficial – a mountain stream or river perhaps. Don't forget to set your intentions. Ask for healing, nurturing, safety, nourishment and ancestral knowledge. Grab any opportunity to wild swim if the weather allows, or take a ritual bath. Pampering spa day with your best friend or partner instead of sitting in front of a screen? Do it!

RECIPE FOR CANCER MOON BATH

Ingredients
- ☽ ½ –1 cup Epsom salts
- ☽ A few drops each of lavender and rosemary oil
- ☽ A rose quartz or moonstone

Method
1 Add everything to warm bath water, including the crystal.
2 Light some candles and play some soothing music.
3 Stay in the bath for 15–20 minutes and just breathe.
4 Let yourself float.
5 When you have finished, watch as the water drains away. Visualize old energy draining away and being absorbed by the Earth.

New Moon in Cancer

New Moon in Cancer is awakening the story of our childhood and how we were nourished and nurtured. It will remind us to prioritize healing and receiving – physical, emotional and spiritual.

We can give ourselves what we might have lacked in childhood. We can create our own home. We can become our own mother – nurturing, compassionate, giving, but also assertive and protective if need be. We can be our own father – practical, resilient, persistent, systematic, taking care of the physical needs, creating structures.

Build confidence in your own sense of self and move through the discomfort. This is a pivotal point in your personal growth. It makes you choose between the familiar ways from the past and the new truth that is emerging from your soul. New Moon in Cancer is not giving dishonesty a chance. This is when you're developing your inner strength.

Modern ritual – use your memories as a grounding tool

A friend and inspiring women's-health specialist once reminded me of this pearl of wisdom: 'we can use our feelings as portals for grounding.' Visualize the strength, love, groundedness and empowerment you felt in a certain place or situation, or with a certain person – as many times a day as you need.

Open your mind to how 'feeling safe and grounded' should look. For example, when I take myself back down my memory lane to Spirit Weavers, the annual women's gathering in the sacred forest in Oregon, I re-experience the powerful connection and sense of safety that I felt while sleeping under the trees and sharing a space with 800 other women. We all have moments in life that are incredibly powerful and that we can save in our 'memory bank' for future reference.

Full Moon in Cancer

Full Moon in Cancer gives us powerful feelers, making us aware of everything going on within and around us and asks us to make sure we can retreat to our protective container filled with calmness and unconditional love. By respecting your boundaries and making sure they are respected by others, Cancer Full Moon will help you to create a life in which you can fully focus on the tasks ahead of you.

Developing discipline around your needs means being clear about what your body, soul and spirit require for feeling at peace and nourished, and making sure those needs are articulated and adhered to. If it triggers some raised eyebrows, let it. Everyone has their own journey. It is about committing to following your inner guidance towards prosperity and growth in all areas of your life and letting go of people and situations that keep you small.

Modern ritual – try a self-love meditation with rose quartz

Support and ground yourself with rose quartz meditation. This beautiful mineral of unconditional self-love and forgiveness will be invaluable for the Cancer Moon days. Either hold it in your left palm, place it on your body or keep it nearby throughout the meditation.

Visualize yourself being fully in a soft, pink bubble, surrounded by the colour of rose quartz. Feel your heart, knowing that the love and kindness that you give yourself can never be taken away. We all need home, safety, nourishment – anchor yourself in your own being, let this be your mother, the one who takes care of you. When your heart is your home, there's nothing to be afraid of, nothing to fear. You are always unconditionally loved. Each time you breathe out, imagine the air flowing out through the front and back of your heart. Visualize yourself in a bubble of pink light. Enjoy for five minutes.

Ask for
guidance
and healing

MOON MAPPING

Moon in Leo

Moon in Leo days are like royal sunshine, focusing on our creativity and bold expression, grace and courage. Leo's courage is deeply rooted in the heart space, full of love for oneself and all there is, which is different from the more impulsive Aries energy. We are all here with the purpose to channel our magical light and creative gifts, unique to each of us, with playfulness and seriousness in equal measure.

When Moon is in Leo, don't be afraid to shine and get noticed. Send out the radiant sunshine from within, from your heart. Dress in bright colours or white and gold; wear bold jewellery. Leo energy is definitely not that of the childish clown, but deeply regal, his appearance representing the internal state and life's purpose, symbolizing your essence and who you truly are. New haircut or shopping trip – yes, please! Reading your poems out loud – absolutely! The point is to do anything that pushes you to express your creativity. Expand your self-imposed limits and boundaries and let your light shine.

Leo energy teaches us to have the powerful presence of one who knows his identity intimately, and achieves more by doing less. Leo encourages us to learn to protect our precious time and energy with a fierce roar if needs be. He only gives in a way he chooses, when he chooses, and you'll know when he does, as it will change lives. He is discerning and doesn't waste energy – he knows that wherever he channels his power, reality shifts, so he chooses wisely.

Reign over your domain with grace and assertiveness. Let go, once and for all, of anything that is not helpful in your path. Leo's priority is engaging with quality – 'golden' quality in everything and everyone you surround yourself with.

Invitation – what to explore?
- Do you connect to your core, your centre daily?
- How do you express your creative side?
- Is your current wardrobe really expressing who you are?
- Do you protect your energy and time, by politely saying No to requests and invitations that are not aligned to your being?
- Could you create more time and space for self-exploration and self-expression?

Moon energy – masculine fire
Leo symbolizes sovereignty and agency and invites you to govern your domain with inner knowledge, wisdom and grace, like a wise king or queen. Precise in words and action, the manifestation

of mature Leo energy is being the master of your own mind and body, and leading your life with compassionate discernment. As creativity is mostly linked to the feminine energy, the masculine aspect of Leo means that the creativity here is deeply focused (knowing who you are and what you're particularly good at), expressive and outward-oriented.

Whatever is sabotaging your creativity, growth and gifts – let it 'burn' with the fire of passion. Take the centre stage, the driver's seat of your life. Roar from your heart. The world has been waiting.

How to benefit – develop self-mastery

Leo is representing the passion for self-mastery. Use the essence of the lion/lioness as your inspiration: calm, centred, royal and monarch of your inner kingdom. Walk and talk with grace and power – remember, the energy you give out matters more than words.

Generally it's a good time to schedule any important meetings – job interview, first date, pitch to a client, public speaking – for the day when Moon transits Leo. Practise going slow and grounding in your royal essence before, perhaps through meditation or breathwork. I always find that I feel calmer when I connect with the people in a room by looking them in the eye (like a queen) during any presentation, rather than staring at a point on the wall or the screen. Go slower than usual. Keep checking in and connecting to your breath and body, as well as to the ground beneath you, throughout. The king or queen never grasps or hustles, just responds.

Tools

Citrine is my favourite **crystal** to work with during Leo days. It shines its yellow light like a Sun and connects me with my joy. You can use it during meditation, by holding it in your left palm and breathing in the vibration of this yellow mineral, or you can simply have it near you while you work or create.

Burning candles or making a **fire** if possible is another great way to work with the fiery days of Leo. Fire has so many different properties – cleansing, transforming, igniting. Try switching the lights off when it's dark and using candlelight instead.

New Moon in Leo

Vulnerability is a strength. The lesson of this New Moon is to soften and let go of pushing yourself. Stop trying to get something quicker or somewhere earlier than you are meant to, without gaining experience and walking your walk. Everything has a divine timing.

When we soften, we align with the rhythms of the Universe and our own truth. When we become vulnerable, we connect and relate to each other. When we feel gratitude, we see how much goodness, beauty and good fortune there is around us already.

Use this time to generate self-love towards yourself, and gently invite creativity back into your life. The best way to do this is embody it, to go by feeling. When we are busy in our everyday lives, we often forget what it means to play or just soak in the warmth of the Sun and be.

Leo New Moon is like a ray of light, showing you the way to your inner strength and lighting up the parts that need to be nurtured and cared for in silence before being expressed. Temporarily pulling back in order to change the way you show yourself to the world is a powerful sign of wisdom and inner strength.

Modern ritual – let sunshine into your heart

Leo governs the heart and is governed by the Sun. This is a beautiful meditation to expand your heart, so that it radiates light:

- Sit or lie down.
- Think of something that puts a smile on your face (maybe it's your child, a puppy or your favourite food).
- Take a deep breath and let your heart and chest expand with joy, gratitude and tenderness.
- Exhale, keeping the chest expanded.
- Inhale again and expand your chest even more.
- Repeat for a few more breaths.

When you finish, sit and let warm, loving and joyful energy fill your whole body.

Full Moon in Leo

This Full Moon has certainty and powerful agency like nothing else. It acts with kindness, but utter precision, boldly cutting ties when and where needed in order to roar freely (aka expressing your truth).

The invitation is to ground with graceful fierceness, set clear intentions to strengthen and protect your energy and physical body. Observe and get clear about which people and situations leak your energy, so you can set necessary boundaries with the waning Moon and create space for playing, loving and new things, people and experiences to fill your life.

Change your game, welcome new directions, new opportunities, new clients, new ideas and new habits that are aligned with who you are. Knowing how to let your true soul shine may just redefine the whole way you live and work.

Modern ritual – dress up

Be seen. Let your inner artist and eccentric shine. During this Full Moon go through your wardrobe and dig out long-lost pieces that you love, and supplement with some new bold designs (supporting ethical brands if possible). Or invest in a magnificent piece of jewellery, or some lacy lingerie. Anything that is out of your 'ordinary' look.

Let your light shine

Moon in Virgo

Imagine the time of year when the whirlwind of activities is over, and you're ready to slow down and focus. Go slower and with intention. Carefully sift through your to-do lists, dividing the tasks into what has to be completed and what needs to be replanted or nurtured.

Virgo Moon lights up our focus to detail and ability to analyse pros and cons with a practical and objective eye. How resilient are the seeds you've planted? This Moon can provide immense courage and discernment to choose what's right for you, live by your own justice and moral code, walk your path in the way you see fit, as service for yourself and others.

But for this, we need to come into full integrity and know our light and shadow as intimately as we know other people's (Virgos observe and analyse every detail). This is the Moon energy you can't fool. If you want to overhaul your health or career, start making a plan here.

I deliberately didn't mention relationships, as Virgo Moon is objective and wouldn't let us dive into any relationship drama. Instead, you might get the urge to sign up as a volunteer, perhaps at an urban garden or farm or in a retirement home. Virgo is also the great healer, but not in

any mysterious way. She is deeply connected with Earth as a wise medicine woman/man. Remember, you're becoming your best self for being able to share your gifts with the world and Mother Earth.

Invitation – what to explore?

- ☽ Are you ready to define what you no longer need or want with a practical and objective eye?
- ☽ Can you define what's yours to fix and what's not in your life?
- ☽ Do you feel any fear around being fully self-sufficient and independent?
- ☽ How could you learn more from nature?
- ☽ Could you share your free time and your energy in a valuable and charitable way?

Moon energy – feminine Earth

Virgo Moon is like a divine Earth Goddess with strength and power that can move mountains. She is an independent woman, who stands on her own, feet planted in the soil. 'Virgo' in Latin means 'self-contained' rather than the image of her as a gentle fairy or 'virgin' that has been created by the patriarchy. Virgo Moon, hence, invites us to find our wisdom, power and self-sufficiency from connecting to the Earth, rather than being dependent on other people and relationships.

How to benefit – make your list and tick away

Remember that you are whole as you are. Virgo asks us to be of service, but it has to come from a grounded space. Not from a sense of lack and need for validation, but from the place of sovereignty, empowerment and true embodiment.

Connecting with the power from within is what this Moon is asking us to do. Find the healer within through rebuilding your connection with nature. Embrace the dance of light and dark. Deep inside, we all know that we are perfect and imperfect simultaneously.

Virgo Moon loves ticking boxes. Getting stuff done. Tidying up the sock drawer. So make a plan and streamline your life. Know that every walk has to be walked, so get to know your own ways – strengths and weaknesses – intimately. Write them down. Make lists.

You are not a martyr, but a king/queen empowered by your knowledge of all aspects of life.

Tools

Know the importance of self-care and wellbeing that come from connecting daily with Mother Earth in whichever way possible – even if it's having a glimpse on a cobweb in the sunlight for just a moment.

Two of my favourite ways to connect with **nature** are walking barefoot in a forest or park, or just taking my shoes off when sitting or standing on grass. Get acquainted with herbs – there are so many beautiful ones that are readily available and can really boost our health. Although I love all the adaptogens – natural ingredients that help the body adapt to stress – that come from far away, we can also look to our own gardens and make friends with dandelions or lavender.

New Moon in Virgo

During Virgo New Moon we need to focus on our own healing, as this is the only way we can truly serve others from a place of wholeness. This Moon has a close, analytic, look-at-your-growth strategy, like an accountant or investor, to see where you are spending in excess and where you are in debt, what makes you blossom and what brings drought.

As Virgo is so closely connected with the Earth, this New Moon is a perfect time to find the connection through simple everyday activities and make small changes, a step at a time. Virgo Moon gives us the power and energy to keep evolving.

Modern ritual – make a meal (and eat it mindfully)

Food is a powerful medicine and we should treat it with reverence. Preparing a meal and mindful eating are practices that easily help us ground and connect with nature. Personally, I am quite reluctant to cook, especially as my husband can outdo me any time with his culinary creativity and skills. But, every time I make an effort, something shifts. I'm not talking about spending hours in the kitchen – even making a quick and easy batch of organic kale chips is a start!

Full Moon in Virgo

This Full Moon has the eye of a hawk, discerning and knowing what works and what doesn't. She's a meticulous one and doesn't let you off the hook. You can't keep telling fluffy stories forever as Virgo Moon knows exactly what's real. Nothing escapes her. The task here is to bring things that are falling away to completion. Go back and finalize anything you've run away from. Return to the details.

Virgo Full Moon illuminates the empowerment from within when you untangle your past (with objectivity). It brings focus and perseverance to deal with past wounds and stories, so you can rise and open up to the wisdom of life in a completely new way, receiving guidance from your body.

Have you tied up all the loose ends? Have you bypassed a completion to get to the next step quicker? Have you sealed the envelope on the parts of your past that still need to be worked through and finalized with grace? Remember to go easy, as Virgo can be quite a perfectionist and the harshest self-critic!

Modern ritual – write a letter of gratitude

Write a letter of gratitude to the city you called home, to the house you lived and loved in, to the relationships you were part of. They are your biggest teachers. Don't run blindly towards your future before spending the time needed to close a door on your past.

'Dear _____ , I want to say thank you for being such a great place/person for the past ____ years...'

Know your
light
and shadow

MOON MAPPING

Moon in Libra

Libra Moon strives for peace and beauty, it desires equality and justice. Libra is the peaceful warrior, the one who fights for the rights of all with love and compassion, not anger. Libra Moon days can be wonderful and regenerative, balancing and soothing. This is the time to be gentle and cleanse yourself of toxicity and darkness. Look at everything with the eyes of harmony and ask how can YOU bring more of it into the world.

Libra loves partnerships and reminds you that there's no need to go it alone – we are not living our lives in a vacuum. We are not fighting our battles alone. How can your relationships, partnerships, friends, family or community support your purpose and actions? How can your purpose and actions support your friends, family and community?

During these Moon days we dance with the relationship between harmony within and harmony with others. We are here to give our light and gifts to humanity, all beings and the Earth, but at the same time remember that the most important relationship is the one with yourself. Every little step towards a life filled with balance, love, deep listening and beauty affects your work in this world in a big way. In fact, by creating harmony and beauty within, you radiate calm and balanced energy outwards, which creates a ripple effect around you.

Invitation – what to explore?

- ☽ Is your work bringing more peace, justice, beauty or harmony to the world? How could you amplify that?
- ☽ What would happen if you spent more time with people who are completely at peace with themselves?
- ☽ Do you receive as much as you give or vice versa?
- ☽ Which small steps could you take to create more beauty in your space?
- ☽ What would happen if you were to shift from complaining to a permanent sense of gratitude?

Moon energy – masculine air

Every zodiac sign has a 'ruler' planet, and the one that governs Libra is Venus. Venus is actually categorized as feminine, which explains the Libra symbol of scales, of balance. It is affected simultaneously by both polarities, masculine as well as feminine, and asks us to find the balance between the two within us and in all our relationships – to find a way to balance structure with flexibility, dreams and inspiration with actual action, and giving with receiving.

How to benefit – balance giving and receiving

Observe and redefine the volume and quality of giving and receiving you are allowing, or blocking. Contrary to the common belief, that feminine = giving, the characteristic of feminine is actually to receive and create new from what has been received, while the purpose of the masculine is to hold a container for this creation and channel it to the right place. Are your feminine and masculine in balance?

If your output exceeds what you receive, it depletes you of your own power, which means you need to practise receiving – money, help, healing, compliments. At the same time, if you always receive but withhold giving, ask whether there's a deep-seated fear that you cannot replenish what you give, that there's not enough to go around. Practise giving in the form of selfless service – donating money to a charity, helping a friend move house or giving an elderly neighbour a hand with the gardening.

Tools

Essential oils have powerful healing energies. My favourite for balancing the heart energy during Libra Moon is **rose essential oil.** You can mix 1–2 drops with some jojoba or another carrier oil and either rub the oils into your sternum, the location of the heart energy centre, or use it as a perfume.

I'm a huge fan of **cacao** and **cacao ceremonies** and this plant is big in the harmony department. Cacao is a powerful plant medicine that contains theobromine and monoamine oxidase inhibitors, which have mood-enhancing and heart-opening properties. Cacao can really be called nature's antidepressant, creating a feeling of connection. And, on top of this, cacao contains more antioxidants per 100g than any other superfood. You might want to try out making tonics with ceremonial cacao yourself – there are some good sources you can use for this, such as the Cacao Club (see page 157 for details).

New Moon in Libra

Libra New Moon wants you to have a beautiful and peaceful life, filled with harmonious relationships, in a magical place called Earth. We can all be our best selves in peace and solitude. We can find calmness in routine, solace in nature and connection with ourselves when we carve out enough solo-time (although hanging out with yourself can also be challenging), but this is not always how our lives play out.

This New Moon asks you to call back your strength and power in relationships – either personal or business. To practise receiving, trusting and asking for equal and just exchange. However, as with all relationships, it takes two to tango – so make sure that you take personal responsibility for your part before demanding that others take responsibility for theirs.

Modern ritual – pull your weight

Choose a relationship that's important to you: with a partner, a close friend or a business partnership.

Contemplate whether there's something that has somehow gone stale, that lacks gratitude and love, has became a routine or perhaps even creates resentment. Consider these questions:

- ⟩ Are there ways you have held back love and compassion?
- ⟩ Have you been distant?
- ⟩ Have you been 100 per cent honest about your own boundaries and needs?

Take full responsibility for your own part. Choose to change your part. Give gratitude.

Ask about and listen to what the other is going through and offer them support. Do something nice for them without expecting anything in return. When someone seems irritated, it's often not about you – rather, it is their cry for some support, some understanding and some compassion.

Full Moon in Libra

Libra Full Moon is especially great for renegotiating relationships and forging new ones, with the goal of moving towards harmony and balance in the relationship dynamic as well as creating more beauty in the world as a team or partnership. For example, I married my husband during Libra Full Moon.

This Moon reveals who is here to help you to manifest your dreams and to guide you on your journey: who will help you to reach your star. As much as we believe in independence and self-sufficiency, we also need to embrace and practise our receiving skills.

Libra is a mind-based air sign, so we should explore giving and receiving with non-attachment and objectivity, not as emotional cries or outbursts. Have conversations, create boundaries and new agreements if needed, with a calm mind and open heart.

Modern ritual – practise giving and receiving

Schedule a (monthly) meeting with your best friend or romantic or business partner. Take a piece of paper and write down five things you receive from the other person: how they nourish, inspire or support your life or work. Start each point with 'I'm grateful for...'

Read your list out loud to them.

Next, write down five things that you'd like to receive and why. For example: I'd love to receive some flowers every once in a while as they'd really brighten up the living room; I'd like to receive a massage as touch is important for me; I'd like to receive support with my business plan as I'm feeling stuck. Ask if the other person is willing and able to give you what you'd love to receive and then set a specific time – for example, could you help me with my business plan next Tuesday?

This exercise helps you to develop mastery in collaboration, delegation, empathy, leadership and asking for help when needed.

There's no need to go it alone

MOON MAPPING

Moon in Scorpio

Scorpio digs deep. It doesn't stop before everything wounded and 'unhealthy' is brought to the surface for the *grand finale*: release and transformation. Scorpio Moon can be easy and incredibly insightful, bringing relief to age-old problems. And it can be challenging and deep. But, overall, these are the days that help you to shift things at soul level. Loosen the grip and let this powerful Moon bring healing and growth.

The process of purging is mostly not much fun while it is going on. I bet not a lot of us are super excited at the thought of spring cleaning or a major clear-out. However, once it's done, we feel lighter, able to breathe deeply and be more ourselves and are left only with things that are truly us and bring us joy, rather than things that deplete our systems with junk.

The alchemical transformation is not always a harmonious matter, not always sane nor filled with comfortable, quiet conversations. The process of transformation can be fierce and loud and chaotic. It can be painful and confusing, but if we are able to stay aware and conscious of what is happening and see the root cause

of our upset, upheaval and pain, we are taking a big step closer to our truth, pleasure and happiness. The secret is to learn to flow between chaos and order, known and unknown, consciousness and unconsciousness.

The main reason we keep ourselves stuck in the mud is that this is what we think we deserve, this is what feels familiar. But you are free to swim to the surface for a nourishing breath and to feel the sunlight. You are free to swim toward what enriches and expands you. If your heart wants to contract, open it up even more. When you expand your ability to receive love, opportunities, kindness and abundance, the old will fall away effortlessly.

Invitation – what to explore?

- ☽ What are the stories of your ancestors, your past lives?
- ☽ Where do you need to go to heal your connection with your roots?
- ☽ Do you feel truly empowered in your relationships and career?
- ☽ Is there a deep trauma you need to look at, heal and release during this Moon?
- ☽ Who are the people – therapists, healers and so on – who could help you to dive deeper?

Moon energy – feminine water

Scorpio often brings the suppressed emotions and feelings, such as rage, to the surface from the deep undercurrents of our soul, and makes us want to 'destroy' everything to get rid of it. As the fierce feminine/goddess energy has been suppressed for a long time, it's not surprising that now it bursts out like a volcano, asking to be witnessed and healed. What is needed here is the calming masculine energy. This acts like solid riverbanks to contain and channel the feminine energy in a healthy way. In a way that creates positive change and makes us voice our needs and feelings, not just blindly lash out in a rage.

A good illustration of Scorpio energy and our needs can be seen in the story of Kali, the Hindu goddess, who was so full of rage on the battlefield that she wanted to destroy the entire Universe. Her male partner, Shiva, calmed her by lying down under her foot, thus channelling the energy, and simultaneously receiving her blessing.

The feminine creative energy within each of us is growing fiercer, the repressed emotions are surfacing and we need our voices to be heard and our bodies to be seen. Water can be calm like a perfectly still lake, and it can be destructive, like a roaring tsunami. In order to calm and use this bubbling creative power for healthy and positive change, we need the masculine energy (within each of us) to channel it.

How to benefit – step up, speak up

Feminine energy is comfortable with chaos and darkness. Although it has been diminished throughout the last centuries, we are now reminded to embrace the shadow and allow the ancient wisdom and memories to come through. We might feel shy at first, not knowing how to live our truth or express our creative gifts, but we just need to step up and speak up. There is great wisdom that needs to be revealed. Lean in.

Tools

Scorpio energy is about the hidden, about the parts of life that have been shunned or manipulated into an unhealthy expression. Sex, money, power – this is what Scorpio Moon asks us to explore. Below are two aspects, or tools, that can empower our lives and businesses more than you think:

Sacred sexuality – When we cultivate healthy sexual energy we live empowered lives. Sexual energy is our main energy for manifestation, not just babies, but anything we want to create.

Money is energy – We need to step into our self-worth to ask for a just exchange of money (or some other energy) and our time, energy and gifts.

I've included some further reading for both of these tools in the References list on page 157.

New Moon in Scorpio

This New Moon is nudging us towards deeper healing. Scorpio is knocking on our door to remind us that we can't postpone living our purpose or push away the gifts life wants to give us. We are asked to dig in the mud and make (inner) decisions about each aspect of our lives, with compassion. This is the opportunity to let the parts that are not supporting our soul and innermost vision of the future die away.

Scorpio New Moon is rebirthing us into new beings. It asks us to be OK with our shadow, be OK with what we've experienced in the past, the choices we made. We have the opportunity to say No now, to do things differently, but often we keep dwelling on the old, feeling that we are forever tainted, forever marked. What is most helpful with this New Moon is to see how everything in your life can be turned into a strength, a gift, your unique experience that rises up from the ashes like a phoenix. Let's share the stories of wounds that we have carried within, repressed, in order to keep peace and harmony within and without.

Modern ritual – list your deepest secrets

Our bodies and cells are the storehouses of all the energy and trauma we have experienced. In our society, we don't often talk openly and share the stories of our past (if they are something we feel ashamed or embarrassed about), hence keeping everyone imprisoned in their shadow without making sense of our journeys and bringing it into consciousness.

A beautiful and deeply transformative ritual is to write down a list of everything you feel ashamed about, that you're keeping secret, the past experiences that still hurt or embarrass you, things you wish you'd done differently.

Now, next to these, write down whether any of these experiences has made you stronger, taught you a valuable lesson, made you more humble or open, shown you your life purpose.

If you feel ready, you can share it with a close and trusted friend or therapist.

Full Moon in Scorpio

Going through Scorpio energy is not a small feat. This Full Moon isn't something that just floats by without stinging. But it holds the power of greatest transformation compared with any other time. It digs deeper than anything else and supports us with releasing ancient traumas on a cellular level.

Strong emotions may help us to connect with our bodies as they are often deeply visceral. If we can feel pain, we can feel pleasure and vice-versa. Scorpio invites us to feel the depths, rather than being numb and disconnected to the experience of life.

It doesn't matter whether your experience is a subtle one or manifests as a huge energetic rupture in your body, the key is to know that you don't have to know. Scorpio is the unconscious: your only task is to allow anything that's emerging from the depths to be felt and processed, with faith in your body wisdom and surrendering to the works of the Universe. Trust that there's wisdom that knows the alchemy of healing better than your mind and willpower.

Although Scorpio can be quite intense, it is the carrier of our biggest transformation. Whatever comes up, Scorpio Full Moon illuminates, burns and transforms, so we can start a new day with more clarity, lightness and a renewed sense of self.

Modern ritual – make space

You need space to heal and evolve. Space for diving deep into your soul.

Make your home a sanctuary, rather than a place where you're a maid, plumber and butler. Delegate everything you don't need to personally tackle. Reduce your domestic to-do list to channel your energy into things you enjoy – perhaps arranging flowers or writing a book. Focus on what's important to you. Get a cleaner or ask your partner or children to help you with domestic chores. Get a babysitter. Get a dogwalker. Use the time gained to dive deeper into your personal experience, your feelings – or have more sex with your lover or partner.

Learn
to flow
between
chaos and
order

Moon in Sagittarius

Sagittarius Moon brings hope, new ideas and big dreams. This energy is a mix of fun and determination, as you need both for ultimate success in whatever you're creating. Sagittarius is all about learning, long-haul travel and adventures. It's also about hope and freedom. The new. The exciting. Rewired connections. Sagittarius wants to pursue his/her dreams and feels resentment towards being locked down or limited in any way.

Being half-man, half-horse, shooting an arrow towards the skies, Sagittarius will ask you to make sure you know your target, and fully discern the difference between your ultimate purpose and the stories that are injected into your mind by society or the media. True freedom requires a solid ground. We need a strong inner core in order to see the future and know that it will manifest as long as our eyes are locked on the star, the horizon. Sagittarius reminds us that we are truly one with everything around us and living a magical life. And we can take leaps, knowing that we'll be held and supported.

As Sagittarius is half-human, half-animal, it reminds us to drop into the natural world for answers, ask questions of the trees, the stones, the animal spirits, and not to forget to consult our own heart. The answers are already there. We can embrace the external or temporary limitations, restrictions and challenges (with love) and yet be able to connect with the bigger picture and inner guidance.

Plant your feet on the solid rock in the Earthly world and know that, at the same time, we are one with the stars.

Invitation – what to explore?

-) What is your heart yearning to learn and do?
-) Where do you want to be in 5–10 years?
-) What are the main challenges on your path to your envisioned future?
-) Are there ways to solidify your foundation for taking off – perhaps saving money so you can leave your job or learning a new skill?
-) Are you on YOUR path?

Moon energy – masculine fire

Yes to outward energy. Sagittarius is definitely a doer. It's the fire of ignition. Fuel your curiosity and inspiration. Find your mentors. Invest in experiences, courses and workshops.

This is such an enjoyable energy of play and hope, like the happy flicker of a bonfire on midsummer night. You know that the new day is starting, the new dawn is on the way and you are ready for action.

How to benefit – find your *ikigai*

This Moon cycle demands honesty. Seeing clearly what you're doing, how you're presenting yourself, what you are filling your precious time with, and comparing it to your internal truth.

Ikigai is a Japanese term for 'a reason for being' and it is exactly what Sagittarius Moon asks us to get clear about before we tie our shoelaces and get going. To reach for your star, you need to be precise in directing your (Sagittarius) arrow. You are more powerful than you think and whatever you choose to become will be.

The most important question now is: are you're going in the right direction, or what's your *ikigai*, your reason for being?

Tools

The best tools for Sagittarius are **exploration** and **travel.** Sagittarius energy is inviting us to be students again. The newbies. Permanent learners. It's looking inwards and outwards with new eyes and developing your ability to be inspired and inspiring.

Become curious about embodied knowledge – the knowledge that you've come to through experience – and your own roots. In order to actually make use of it on a deeper level we need to experience it – either by travelling to places with different cultures or by exploring the traditions of our own ancestors. Perhaps you could visit sacred sites that would once have played an important role in their lives, or try using the plants and herbs that were important to them, or you could try celebrating a ritual that comes from your own tradition. You can travel near – by exploring your own surroundings, your current homeland – or far.

If you are flexible about where you can work, why not pack your laptop and head to a different country for a week or two?

Start planning your next work trip when the Moon is in Sagittarius.

New Moon in Sagittarius

This New Moon should be full of nurture and enjoyment. Take a break and be grateful for all the joys and lessons, all the huge changes that have occurred, and start visualizing your new life. Sagittarius makes our focus clearer. It helps us to choose a specific star in the sky and direct our arrow towards it.

Sagittarius New Moon marks a moment of great expansion of our minds, magical connections with human and spirit guides and the possibility of waking up to a very different existence.

Modern ritual – visualize your future

This one puts a smile on my face every time, no matter what's happening. It's great for the moments when you feel worried about something or challenged in any way, or when the heaviness of the world weighs upon you.

First of all, replace worry with gratitude. Somebody reminded me that if we put the same effort and ability to imagine that we use for worrying into feeling gratitude for what is instead, life would be transformed. So close your eyes and feel everything you are grateful for.

Then imagine your life in a few years. Visualize where you would live (as specifically as possible), how you would work, who you would spend your time with – and, most important, really FEEL it. Feel the joy of living the life of your own truth, empowered and nourished, giving and receiving.

Do it daily, if possible – enjoy it and make it happen!

Full Moon in Sagittarius

Sagittarius Full Moon is like a message from your future self. It might come as a sensation or feeling of expansion, or knowing. Or maybe it manifested as synchronicities or funny little things such as your grocery bill coming to £11.11. In whatever way it occurs, it (hopefully) gives you some comfort while you are in the 'in-between space': knowing that your old skin is being shed and the new one is still fragile, vulnerable and translucent, and it will take some time before you feel totally comfortable in it.

Modern ritual – reveal what the 'being' is behind your work

Sagittarius energy is one of lightness, joy and expansion. If you're not enjoying and loving what you do, it's time to take a step back and write down what the 'BEING' behind your work is.

When I did this exercise during the Full Moon in Sagittarius, it blew my mind and opened my eyes. While Mylky Moon Lab was all about 'being receptive, nourishing, being in process, being feminine', my digital marketing consultancy was 'being rigid, masculine and results-driven' (my body literally felt tense). Although I love and feel nourished by the collaborative aspect of my work with social media, this gave me the necessary push to completely change my business model and interweave the two 'beings' together for a united vision.

Know your target

Moon in Capricorn

Capricorn energy is one of ultimate persistence and determination in a very practical and real way. Not fazed by challenges, she will give us the courage and strength to keep going even during times of confusion, when things get tough, or fear kicks in.

When you start tracking and observing the character of the Moon on different days, it might feel that Capricorn has a somewhat serious quality. There are mornings when I wake up and feel the 'grind', the pressure of getting to my work and to-do lists as soon as possible, and these mostly happen when the Moon has moved into Capricorn's part of the sky.

Although extremely structured and hard working, Capricorn energy actually has a deeply nurturing and mothering aspect. Capricorn never says No to lending a helping hand or supporting someone in need. However, what might be the challenge at this time is nourishing YOURSELF through daily rituals and routines that support your resilience, such as morning meditation, a healthy tonic or regular yoga practice.

The good thing about Capricorn energy is that there is no giving up. Ever. Even when things seem hopeless, even if you fall many times or make countless mistakes. You will rise and walk on – like a single mother who takes care of her children while working full time. She keeps climbing upwards, even amid the avalanche. So don't get disheartened when you fall off the wagon time and time again, just climb back on and keep riding.

Everything is in cosmic order: the Sagittarius energy that comes before Capricorn helps us to see our destination, and Capricorn days help us to manifest these visions. This Moon gives us the necessary grit and determination to build our dream, one brick at a time, while we soften and slow down our step, and cultivate patience and perseverance. Just make sure you provide yourself with all the nourishment you might need for this long and magical journey.

Invitation – what to explore?

》 What are your top three priorities this week, month and year?
》 What are your biggest fears?
》 Do you have specific daily rituals and routines that nourish you?
》 Do you reward yourself when you accomplish something?
》 How much of your day or week do you block out for self-care and personal development?

Moon energy – feminine Earth

Capricorn energy is practical and grounded. Not one who blindly runs for whatever carrot is dangled in front of her, she's committed in her journey and knows the destination – the top of the mountain.

Capricorn asks you to amp up the discipline, honesty, compassion and gentleness towards yourself and others in order to support the path you have chosen. Build the strength of your heart and body, not just your mind. Nurture yourself through steady routines and rituals. The more you can connect with Mother Earth, the better.

How to benefit – start a spreadsheet

Having many ideas and interests is great, but make sure there will be a box to tick at the end. Capricorn is your ally in that department.

Once you have a clear idea of the path you are here to walk, it's time to lace up your hiking boots and get going. Capricorn is the master of lists and structure, so keep your focus on the point you're planning to reach and start creating your list or mind-map of the steps, milestones and resources needed. This Moon loves nothing more than a proper spreadsheet.

Tools

One word – **planner**. Whether it's on paper or online, a planner is the best tool for the Capricorn day.

New Moon in Capricorn

Less is more. Quality before quantity. Focus on your purpose. As an Earth sign, Capricorn New Moon grounds us in the Earth reality and gives us the energy to diligently navigate the mountain roads during the next cycle.

Are you are swimming around with lots of ideas and feelings, but struggling to find the focus and determination to make the changes and keep going? This is the best time to set intentions that make you more productive and efficient, while making sure you don't waste your time on low-level priorities.

Capricorn New Moon helps you to grow your capacity for efficiency, hard work, productivity and taking care of the everyday practicalities. This New Moon is not the time for some mystical fairy dance ritual, but for 'making it work' in an effective new way.

Modern ritual – commit

Write down three commitments for every day/week/month/year. What are the three priorities you want to focus on in each time period?

Include tasks that you know you've been putting off for a while, but that need to be done. Things that might be scary, but are deeply necessary for your peace of mind and moving forwards. Go to what scares you the most first. Redefine the things that you feel most resistant to by tuning into your body wisdom and intuition to find out whether it's fear that's talking. Perhaps it's not truly your priority, or maybe the timing is not right.

Full Moon in Capricorn

Capricorn Full Moon illuminates the areas that hold us back from shining our light in the fullest and most joyful way and, hopefully, challenges you to fully commit to the change. It lights up the parts of your life and business where you are limiting your own expansion. Maybe you put in a lot of work, but are not getting the results you are after. Or perhaps you are just not living the life you want to live. Sometimes we also need to unplug ourselves from the illusionary security, look at what we believe we should be doing and develop a bit more selfishness.

This Full Moon wants you to take a VERY honest look at where you're spending/wasting your energy, what (or who) is draining you and where those habits are rooted. This might well be programmed into your system via your family, culture or ancestral lineage, so there might be a need for clarification and resolve in that area.

Modern ritual – understand your energy leaks and hacks
One of the best things I learned was understanding energy leaks and hacks.

Energy leaks are all situations, people, activities, foods, drinks and so on that ultimately leak your energy (and diminish the cognitive clarity you need for pursuing your dreams). The list is entirely individual for each one of us. Just pay attention to how things make you feel and there's your answer.

MY ENERGY LEAKS
- ☽ Loud noise
- ☽ Staying up past 11pm
- ☽ Overdosing on caffeinated drinks
- ☽ Too much screen time
- ☽ Staying indoors most of the day
- ☽ Midnight Instagram/email sessions
- ☽ Multitasking

Energy hacks – pour discipline and determination into restructuring your schedule, so it includes downtime and rest, movement and time outdoors, mindful nourishment and anything else that your body, mind and spirit might need in order to flourish. Listen to your body and turn to the Earth. Again, energy hacks are different for everyone.

MY ENERGY HACKS
- ☽ Solo-time
- ☽ Eight hours' sleep
- ☽ Matcha
- ☽ Wild swimming
- ☽ Reading
- ☽ Naked sunbathing
- ☽ Doing one thing at a time with full presence

Make a list of your own energy leaks and energy hacks.

There is
no giving
up

MOON MAPPING

Moon in Aquarius

Aquarius detaches us from the everyday, distances us from 'the known reality' and makes us spread our wings and open our minds to the bigger, cosmic picture. The goal of this energy is to rocket you towards your ultimate truth through imagination, radical innovation and change. Space exploration? Going to the Moon? That's Aquarius. Impossible? That simply means Im-possible, 'I'm possible' in Aquarius eyes.

We all came here with a mission that is far larger than we think. We are here with a bigger purpose and the responsibility to use our gifts and talents as a contribution to our community or society at large. You are not living your life just for yourself or growing and healing just for your own good – YOUR healing and transformation are needed for future life on Earth.

Business can be non-linear, chaotic and mysterious and Aquarius empowers us to be comfortable with that. To keep our eyes on the horizon, or further. It asks us to do work with our soul passion, mix spirituality with sustainability, profit with purpose, focused vision with co-creation. Forget the rules, traditional roadmaps and milestones, forget how things are done – write your own rules, listen to your own body and your own guidance.

Aquarius Moon opens the portal for this new vision for living, working and creating for a couple of days every month. Technology, innovation and collaboration are the keys here. Moon in Aquarius encourages us to come together to create radical change. We can only get so far alone. After all, it took a proper team to send a spaceship to the Moon.

Invitation – what to explore?
》 What can you do to use your passion for serving on a large scale?
》 Are you creating space for yourself to connect with your inner alchemist and innovator?
》 Are you ready to think outside the box and delete old beliefs and rules to write your own?
》 Is using technology a drain or gain for you?
》 Are you ready to start a group in your community to initiate and create positive change on a local or global scale? What could the focus be?

Moon energy – masculine air
The keywords for Aquarius are Mind + Focused Action. There is nothing incremental about this energy. The old beliefs and world will crumble. In terms of much bigger cycles, we have now moved into what is called the Age of Aquarius, lasting over 2,000 years, meaning that the characteristics of the

new paradigm are visionary, creative, rebellious and free – everything we need for creating a world that is a complete shift from the old system.

Aquarius Moon asks you to go beyond the mundane and master the arts of delegation and focus as the path to success. This masculine energy has clear boundaries, which helps you to have a solid container for creation and innovation. Ultimately, doing less and constantly questioning whether what you do is what you need to do in order to expand IS the way to unite clear purpose and focused action.

How to benefit – use your mind

Aquarius Moon is like a green light for brainstorming and initiating change on a huge scale, especially finding new ways to use technology with intention, so that it will serve our wellbeing, the society and environment.

This is the time for developing ideas and getting acquainted with two concepts – cognitive clarity and strategic laziness. Learn to see the bigger picture in your mind's eye. Say No to 'fluff' – aka the energy drains in your life – like office gossip or spending hours on Netflix or Facebook. Aquarius asks you to master delegation to the point that you don't do anything that YOU don't need to do. One of the gems I've heard is that, if you are a business owner, make sure that you work ON the business, not IN the business. Don't get wrapped up in

things you can delegate or hire people to do, or tasks that don't require your unique skills. Only by getting our heads above water can we see what the weather is doing and where the shore is.

Tools

You don't need to hire a full-time crew to help you. There are many companies and platforms that offer support for small businesses, or **VAs (virtual assistants)** that come with no employment contracts or extra admin for you.

Fluorite is my favourite crystal for Aquarius Moon days. It is called 'stone that gives order to chaos', reminding you to step away from the nitty gritty to see the bigger picture. Fluorite is helpful for coordinating physical and mental abilities, providing clarity and rejuvenation and transforming confusion and frustration into a sense of peace.

Fluorite is a great tool for having on your bedside table. I love working with this mineral first thing in the morning when I'm still in bed. Just lay it on your third eye, heart or lower belly and breathe. Feel the calming vibration. It will gently set you up for the day ahead.

New Moon in Aquarius

We often get hooked on our habits and routines, holding onto them as a safety mechanism. But your ingrained stories might keep you locked in some old version of yourself, limiting expansion.

While committing to daily rituals is amazing, we do need to rock the boat every once in a while and question things. Let your imagination and creativity roam freely and try a different version. Your life is constantly evolving, so it only makes sense for your daily habits and rituals to reflect that.

Aquarius New Moon asks you to look at your own life with objective eyes and see where you could innovate. Are there any technologies or apps that could help you to become a better version of yourself?

Modern ritual – activate deep curiosity

I invite you to explore a question. There's a beautiful practice of deep enquiry that you can do with a partner, but it can also be done alone.

If you are working with a partner, sit opposite each other and ask a question, for instance: 'What is stopping you from living your ideal life?' Set a timer for five minutes in which they can answer. Allow the subconscious to do the talking. Silence at any point is welcome.

The task of the listener is to listen intently with no interruptions. Then switch sides and let your partner ask the same question of you. Switch sides two more times, each time listening deeply with no reply or interruption at any point, holding space for whatever appears. Afterwards, take some time to journal.

If you don't have a partner to practise with, you can also do this with free-writing. Just write the question at the top of a sheet of paper and free-write (let anything that comes up to flow onto the paper without thinking about it) for ten minutes.

Full Moon in Aquarius

Full Moon in Aquarius invites us to gather together. We know deep down in our bones that we can't get far on our own, but that together we are invincible. It encourages us to initiate experiences that pull together masterminds from different backgrounds, to work on a common goal.

This Full Moon has cosmic quality, reminding us that we all have something invaluable to offer, our unique ideas and gifts. We are all teachers and students at the same time. You are never the only one on the journey, the only one with a unique goal, the only one trying to make a difference or asking questions, or not fitting in the box.

Design Thinking is something that comes to mind with the Aquarius Moon: it's a process for creative problem-solving with a human-centred core. It's a way of developing ideas that encourage organizations to focus on the people they're creating for, thereby leading to better-designed products, services and internal processes.

Modern ritual – brainstorm
Initiate a Full Moon brainstorming session (with yourself or your team):

1   If possible, and with a team of people, choose your participants from different modalities and backgrounds for added diversity of thought and ideas.
2   Have a clear structure and timing.
3   Choose a location you haven't been to before, preferably outdoors.
4   Start the meeting with meditation or breath work, so everyone can connect with their own inner wisdom.
5   Surrender the control.
6   Let go of the attachment to find answers or immediate solutions and instead listen deeply. Aquarius loves freedom, and the detachment from outcome gives an immense freedom to dive in with much more openness and trust and with deep faith.
7   Know that solutions will come when you relax into the sense of belonging to a community and/or to the landscape. We don't need to 'figure things out' using our minds alone.

Let go in order to expand.

Detach to connect.

Create your own rules.

Curiosity is the key – if it's all you do, stay curious and ask questions, keep exploring deeper, feel into the complex patterns within yourselves and your surroundings – you'll never lose your way.

Write your own
own
rules

Moon in Pisces

Pisces Moon days are sensitive, both physically and mentally. We are ending our Moon journey here and returning to the end of the cycle where we merge with Universal truth and become one, integrating everything we learned on the way, to be reborn as a 'cosmic baby' into the next cycle, starting with Aries.

Pisces energy brings waves of emotions and feelings we often can't even articulate. Situations trigger these on the subconscious level and it might take a while until we finally see it clearly. Pisces reminds us of the importance of getting intimate with the Universal breath and to flow with the cosmic waters. This is the time for going inwards and playing it solo, if that calls to you.

Pisces energy is the one of the psychic, one who feels everything, whether it's joy or suffering, so being in big groups at this time (at a festival, for example) can be a bit of a challenge to your soul. And as everyone who has Piscean energy imprinted in their cosmic blueprint knows, it's not always an easy existence. Pisces Moon doesn't have boundaries, so the love can be abundant, but so can pain and any other emotion.

Above all, Pisces brings the ultimate knowing that we are free. We need to remember that the Universe is not creating us with a well-used mould. We are not the products of a cosmic factory. We don't have to be a certain way in order to fit in – whether it's our own view of ourselves, or someone else's. We don't need to pretend to be polished individuals, please people or project a certain persona for our Instagram feed.

The Universe asks you to seek your own truth, be messy, go your own way and find your own path. Trust that whatever you're doing, feeling or going through is exactly the path that you need to be on.

Invitation – what to explore?

》 Could you keep opening up and breaking through the walls of protection and defence to heal your deepest wounds with love?
》 Could you immerse yourself in gentleness and compassion, in order to receive and channel true love?
》 Could you become more connected to the Universal unconditional love?
》 Can you create healthy boundaries, so you don't take on others' energy?
》 Do you clean and replenish your energy daily?

Moon energy – feminine water

These past centuries have been hard for Pisces energy to flourish. While dominant masculine productivity and linearity have been the place of worship, the fluid and dreamy Pisces was condemned. This feminine water energy has been feeling out of place, belonging to the cosmos rather than feeling at home on the Earth. Forever disconnected and isolated, prone to escape the overwhelm of emotions within and without, numbing her powers.

But we are on the cusp of the new era and, with the rise of the feminine, with the returning respect towards our intuition and inner eyes, the Piscean energy is embraced. Our distorted waters are calming and merging, like two rivers finally meeting and becoming one. Scientists joining hands with mystics and healers. Fashion with nature. Business with astrology. Pisces energy is the one that peels off the layers of old conditioning and initiates rebirth at soul level.

How to benefit – get to know your own magic

During the Pisces Moon it's essential to carve out time and space to be on one's own, float in the cosmic dream, swim around freely in your own inner lakes. Every once in a while, we must break free and have our own communion with the divine wisdom and all there is. Withdrawing from the external stimulation and allowing inspiration to come from inside.

How many workshops, courses and retreats have you been to this year? Quite a few? Pisces reminds us that consuming the information fed to you will only make sense when you understand that the ultimate power, wisdom and healing come from within. It's already there.

Tools

I love working with **light blue celestite** during Pisces Moon, reminding me of calm waters. Celestite brings a sense of trust and flow, and is a symbol of calmness and peace. When you live and breathe the Universal love, you'll have the mental clarity to resolve anything. It also connects you with your inner guardians and guides, which is necessary for supporting Piscean out-of-this-world existence.

I particularly like to work with celestite before I start writing, even if it's just an email. You can try this, too, it only takes a few minutes. Close your eyes, breathe and hold the mineral in your left palm (the left side is better for receiving energy as it's referred to as our feminine side). Feel the cooling calmness caress your soul and open your creative centre.

Journalling is essential during Pisces Moon – or anything else that gets your creative juices flowing freely, translating words or images directly from the unconsciousness.

New Moon in Pisces

Pisces New Moon has a quiet power. A whisper that you feel in your bones. As the last sign of the zodiac, this Moon is the great mystic, the selfless dreamer and lover, sensitive to all the subtle energies around us. We have peeled off a lot of layers and learned from many insights during the last cycle – so this New Moon is the time to slow down and exhale.

For the Universe, your personas and masks are not important. How many hours you spend at work, the amount of money in your bank account or the car you drive are not of interest. However, your healing is EVERYTHING. The Universe wants you to become a vessel for pure light. To fill yourself up and beam it out to all beings who might need it. Especially Mother Earth.

When you heal, you start receiving. Your energy starts flowing in a different way. You get guidance. Opportunities and people start flowing in miraculously. You'll be supported and held. The replenishing sources of energy become unlimited. We want healing that comes in like waves, no rigidity or pushing. Relinquish the chase for the outcome and let the process unfold at its own pace.

Modern ritual – heal and receive

Imagine that your only task is to heal and receive. Schedule a week of receiving into your calendar this New Moon. Instead of focusing on giving (work, energy, time, support...), amplify what you receive (money, support, nourishing food, wisdom...) from the Universe each moment and see if you can expand it.

Our aim is clearing the blocks that keep us from receiving the magic of the Universe, but also set healthy boundaries. What do you say Yes to that should be No? What do you say No to that should be Yes? What brings you into the flow?

Knowing that healing is your only purpose in this life, above anything else, how would your priorities shift? Which decisions would you make and commit to?

Full Moon in Pisces

Pisces Full Moon invites you to embrace freedom. We often put limitations on ourselves, thinking we are this or have to be that, we must do this or that. In fact, most of what we do is purely our own choice. But the ultimate knowing about who we really are comes from following our feelings. Going deep into our own wisdom and intuition. Not following the path set for us by anyone else – society, media, community or our own ego.

This Full Moon is about remembering. Remembering our intimate connection with the Universe and the divine. Trusting our own path and finding freedom in surrendering to our own flow and dreams.

Modern ritual – create with intuition

The last ritual I want to leave you with is an invitation to inject intuition, body wisdom, alignment, deep listening and flow into everything you do and create – whether it's business, social media content or anything else that might not appear to be deeply spiritual or part of your 'health and wellbeing'.

In fact, Pisces is the portal for integrating all aspects of our selves and not constantly changing masks. You can be as mindful and aware of your body when you are at work as you are during a yoga asana. You can post on social media with the same level of intention and vulnerability as writing a letter to your loved one. Pisces is the eraser of boundaries and stops us from compartmentalizing aspects of ourselves and thus feeling as if we're living many lives at once, and actually get lost in who we really are.

Just be YOU, follow YOUR flow and create anything your heart calls for with full authenticity and ease.

Seek your own truth

Goodbye is just another hello

Usually it's time to say goodbye at this point, but the beauty of cyclical living is that the goodbye is just another hello.

The end is just the beginning of a new cycle. The cycle which will not be like any other that you've experienced in your past, nor will you ever experience the same cycle, in exactly the same way, at any time in the future.

Every moment is precious. Every moment has the potential for change and you never know who you will be by the time the Moon has completed yet another circle around our home planet.

This is the magic of life.

I hope this book will not end up under a pile of newspapers next to your bed or become just another object on the bookshelf. This book is asking to be used, and used again. Read and then reread it once the Moon launches into another journey around us. Make as many notes on the sides as you wish or rip out pages if you need to. Use it as a place to store notecards, plant cuttings, photographs or anything else that inspires you on the journey. This is just the beginning of it all and after a few 'moon-ths' you will feel like the cosmic curtains have parted and you can see that there is a whole Universe ready to welcome you with open arms.

If you'd like to learn more, sign up for bi-monthly Moon Letters on www.mylkymoonlab.com or contact me for a Mentoring Session. I'd love to hear about your experiences during different Moon phases and Moon signs, so feel free to drop me a line whenever you feel called to: mamabear@mylkymoonlab.com

About the author

Merilyn Kesküla-Drummond is a mentor, speaker and entrepreneur living in London with her husband and dog Lune (aka Moondog). Born in Estonia, Merilyn left her home country aged 20 to study Fashion Management in London, which led to her working for several magazines that she had dreamed of contributing to while growing up — including *Vogue* and *ELLE* – as well as travelling internationally as part of her work in the fashion scene.

Finding it hard to settle, Merilyn moved around like an urban nomad – from a summer in San Francisco to an autumn in Monaco – before moving to Zurich, Switzerland, where she stayed for six years working for a Global Strategic Innovation team at a renowned German audio brand. After experiencing almost constant anxiety and bouts of burnout over the years, Merilyn started to explore cycles within and around her and re-ignited her love for astrology. She founded Mylky Moon Lab in 2016 and has hosted workshops around the world, including in Zurich, London, Milan and Copenhagen.

Merilyn also runs a digital marketing agency – Keskula Digital – that applies the Moon Method and principles of Digital Minimalism in its work with social media and content creation.

Website: www.mylkymoonlab.com
Instagram: @mylkymoonlab

References/Useful resources

References

Page 22
The Earth and Moon are isotopically similar
Dauphas, Nicolas, 'The Isotopic Nature of Earth's Accreting Material Through Time', *Nature*, 26 January 2017, DOI org/10.1038/nature20830
Composition of the Moon
https://www.space.com/19582-moon-composition.html

Page 25
Space Race
Parr, Martin, *Space Dogs: The Story of the Celebrated Canine Cosmonauts*, London: Laurence King Publishing, 2019

Page 28
Space Age Fashion
http://www.bbc.com/culture/story/20190312-how-the-1960s-space-age-fashions-changed-what-we-wear
https://www.highsnobiety.com/p/evolution-space-aged-fashion/
https://www.vogue.com/article/pierre-cardin-exhibition-brooklyn-museum-of-art
http://www.vam.ac.uk/content/articles/p/pierre-cardin/
Uniforms for female empowerment
https://www.highsnobiety.com/p/evolution-space-aged-fashion/

Page 31
Osiris in Ancient Egypt
Quirke, Stephen and Spencer, A Jeffrey, *The British Museum Book of Ancient Egypt*, London: The British Museum Press, 1996

Page 36
Lunar gardening
https://www.organiclesson.com/planting-by-the-moon-the-starters-guide/

Page 41
The feminine and the menstrual cycle

http://www.mysticmamma.com/moon-time-and-astrology/
Blackie, Sharon, *If Women Rose Rooted: The Journey to Authenticity and Belonging*, Tewkesbury: September Publishing, 2016
Gray, Miranda, *The Optimized Woman: Using Your Menstrual Cycle to Achieve Success and Fulfillment*, Ropley: O Books, 2009
Owen, Lara, *Her Blood is Gold: Awakening to the Wisdom of Menstruation*, Shaftesbury: Archive Publishing, 2008
Pearce, Lucy H, *Burning Woman*, Shanagarry: Womancraft Publishing, 2016
Vitti, Alisa, *WomanCode: Perfect Your Cycle, Amplify Your Fertility, Supercharge Your Sex Drive and Become a Power Source*, New York: HarperCollins, 2013
Wurlitzer, Sjanie Hugo and Pope, Alexandra, *Wild Power: Discover the Magic of Your Menstrual Cycle and Awaken the Feminine Path to Power*, London: Hay House UK, 2017

Page 42
Moon lodges
http://www.mysticmamma.com/woman-moon-lodge/
https://www.telegraph.co.uk/women/life/why-women-are-gathering-in-red-tents-across-the-uk/
https://www.sevencircles.org/womens-moon time-and-ceremony/
Mugwort
https://www.world-of-lucid-dreaming.com/mugwort-dreams.html
https://www.learnreligions.com/using-mug wort-in-magic-2562031
https://www.fatandthemoon.com/blogs/news/magic-mugwort
Grandmother Moon
https://traditionalnativehealing.com/tag/grandmother-moon

Page 61
Studies looking at the Moon and crime
Thakur, C P and Sharma, D, 'Full moon and

crime', *British Medical Journal*, volume 289, 1984, pp. 1789–91

Wehr, T A, 'Bipolar mood cycles and lunar tidal cycles', *Molecular Psychiatry*, volume 23, 2018 https://www.nature.com/articles/mp2016263, pp. 923–31

Page 63
Frank Brown report on hamsters
Brown, F A Jr, 'Propensity for lunar periodicity in hamsters and its significance for biological clock theories', *Proceedings of the Society for Experimental Biology and Medicine*, volume 120 issue 3, 1 December 1965, pp. 792–97
Study on injuries in pets and fullness of moon
https://www.livescience.com/37928-ways-the-moon-affects-animals.html

Page 68
Study about spiders
https://www.nationalgeographic.com/news/2017/08/animals-react-total-solar-eclipse-august-space-science/

Page 80
Feminine and masculine astrology signs
https://katieturner.love/2015/12/28/masculine-and-feminine-signs-and-gender-in-astrology/
https://stormcestavani.com/2013/10/06/astro-101-the-astrological-signs-part-1-gender-in-the-horoscope/

Page 92
King's College study on nature
https://www.kcl.ac.uk/ioppn/news/records/2018/january/Study-suggests-exposure-to-trees-the-sky-and-birdsong-in-cities-beneficial-for-mental-wellbeing

Page 96
Wellbeing events
https://www.consciouscityguide.com/
https://otherness.co/
https://holisticism.com/facilitators
Podcasts
https://www.calmer-you.com/category/podcast/
https://tobemagnetic.com/expanded-podcast
https://mythicmedicine.love/podcast
https://www.lukestorey.com/lifestylistpodcast

Page 103
Gatherings
https://spiritweaversgathering.com/

Page 120
Essential oils
https://www.aqua-oleum.co.uk/
https://vitruvi.com/
Cacao and cacao ceremonies
https://thecacaoclub.com/
https://www.cacaoceremonies.co.uk/

Page 126
Sacred sexuality
Thomashauer, Regena, *Pussy: A Reclamation*, New York: Hay House, 2016
https://www.grace-hazel.com/
https://www.thejadedoor.com/
Money is energy
Silver, Tosha, *It's Not Your Money: How to Live Fully from Divine Abundance*, New York: Hay House, 2019
Williamson, Marianne, *The Law of Divine Compensation: On Work, Money, and Miracles*, New York: HarperCollins, 2012
https://wearesacred.org/the-3-great-spiritual-lies-about-money/

Useful resources

Websites for Moon wisdom
https://chaninicholas.com/
https://mooncircles.com/

Finding out the current Moon phase and sign
https://www.lunarium.co.uk/
https://mooncalendar.astro-seek.com/
https://www.timeanddate.com/moon/phases/

Moon sign calculators
https://cafeastrology.com/whats-my-moon-sign.html
https://www.lunarium.co.uk/moonsign/calculator.jsp
https://astrostyle.com/learn-astrology/moon-sign/

All websites accessed 22 September 2019

Index

Acknowledgements

I'd like to say a huge Thank You to Olivia Percival and Valeria Huerta, my literary agent, for asking me to embark on this journey. I always knew that one day I'd write a book and they opened up the portal that brought it into reality. Thank you to the publishers – in particular Kate and Polly from Octopus – for being incredibly helpful and supportive, and for believing that Moon is a topic worth exploring by the wider audience.

I'm so grateful for my incredible friends in London, Zurich and everywhere else around the world, who never thought of me as a crazy Moon-lady but have been part of Mylky Moon Lab since its beginnings – even by joining my humble Moon Circles back in the day (you know who you are!).

I'm grateful for Spirit Weavers, which changed my life on such a deep level, Irene Lauretti for answering my Instagram DMs and all the amazing teachers, healers and Moon sisters around me, who have supported my journey. Celine Chappert – you've been a gem and inspiration in every way, and thank you to Elisa Vendramin, Marianne Viktor and Christel Voss who have helped to create the visual identity of Mylky Moon Lab.

I'll be forever indebted to my family in Estonia and London for all the love, support and books about the Moon that I have been gifted at Christmas and on birthdays over the past years.

Most of all, there is something bigger, where all this information came from. So I'd love to thank the stars, Mother Earth, Grandmother Moon, the Sun, the cycles, the air and waters, the plants, the beings and all the magical energies within and around us, as well as our ancestors and guides who are always looking after us and making sure that we are held. I'm sure it was one of them who whispered the Moon stories into my ear.